Cosmic Rays

HARVARD UNIVERSITY PRESS
Cambridge, Massachusetts, and London, England
1989

Cosmic Rays

Michael W. Friedlander

This book is printed on acid-free paper, and its binding materials have been chosen for strength and durability.

Library of Congress Cataloging-in-Publication Data

Friedlander, Michael W.
  Cosmic rays.

  Bibliography: p.
  Includes index.
  1. Cosmic rays.   I. Title.
QC485.F75   1989   539.7′223   88-24371
ISBN 0-674-17458-5 (alk. paper)

Acknowledgments

It is a pleasure to acknowledge the help that I have received, both directly and indirectly, in the preparation of this book. More colleagues than I can list in the cosmic ray community have provided stimulation and friendship over many years. Many of them responded generously to my request for photographs. The manuscript has benefited greatly from the careful scrutiny it received from several reviewers, as well as from the helpful suggestions of Howard Boyer and the excellent editing of Mary Ellen Geer at Harvard University Press. To all of these individuals, and especially to my family for their unfailing interest and support, I am deeply grateful.

Contents

Cosmic Rays

The Early Days

"Coming out of space and incident on the high atmosphere, there is a thin rain of charged particles known as the primary cosmic radiation." With these words, Cecil Powell began his lecture at the ceremonies in Stockholm when accepting the Nobel Prize for Physics in 1950. Powell was being honored for the improvements he had made in a method of detecting high-speed atomic particles and for his discovery among the cosmic rays of mesons, minute but radioactive constituents of matter.

What are cosmic rays? What kind of radiation are they, and why *cosmic*? Where do cosmic rays come from, and what do they do? Are they harmful to us? Now, more than seventy-five years after their discovery, we know that the cosmic rays Powell described consist mostly of high-speed protons, the nuclei of hydrogen atoms, with about 9 percent helium and heavier nuclei and a small percentage of electrons. The Earth is continually bombarded by these particles, which collide with atoms in the stratosphere to produce more particles that can be detected throughout the atmosphere and sometimes deep underground.

We have identified possible sources of this cosmic rain. Supernovas, the explosive late stage in the evolution of stars much more massive than our sun, are known to provide many fast particles, and they seem to meet the requirements for a source of cosmic rays. Most of the cosmic rays probably originate within our Milky Way galaxy, but a very few particles—and only the most energetic—appear to be coming from very much farther away. The general outlines of this picture are coming into a fuzzy focus, but much of the detail remains to be clarified. We know that some cosmic rays come from the sun. We also know that the sun and the Earth's magnetic field combine to exert a major influence on the slowest cosmic rays. Cosmic rays pass through us at all times and give us small but unavoidable doses of radiation.

Any mention of radiation raises public concern today, after Hiroshima and Nagasaki, Three Mile Island, and Chernobyl. This was not always the case, however; in the early years of this century, radiation was seen not as hazardous but rather as something mysterious, even beneficial. Scientific discoveries and research had identified several different but sometimes related forms of radiation, and medical science was beginning to see some almost magical prospects for the use of X-rays in diagnosis and therapy. There was an emotional public response when Madame Curie toured the United States in 1921 to raise funds for the purchase of one gram of radium for her research institute. Arthur Compton won the 1927 Nobel Prize for his research with X-rays, which provided important support for the new quantum theory. Radiation was a hot subject, and people were fascinated by cosmic rays, even though (or perhaps because) they were enigmatic, from somewhere out in the cosmos.

In many ways cosmic rays still retain this aura of mystery. As we will see, cosmic rays have been and remain central to a remarkably diverse range of research studies, from the smallest imaginable scale involving the composition of the atomic nucleus to the vast scale of galactic dimensions. Cosmic rays have been used to probe the magnetic field surrounding the Earth and the regions between the stars. The study of cosmic rays has stimulated the development of techniques for the detection of radiation as well as research into ways in which radiation affects atoms and molecules, in matter both living and inert. Cosmic rays are studied from satellites and space probes, from balloons in the stratosphere, with high-altitude airplanes, and with giant detectors two miles beneath the Earth's surface. The fundamental nature of many discoveries in cosmic ray research has been recognized through the award of a disproportionately large number of Nobel Prizes.

How did all this activity start? The first intimations of the presence of cosmic rays came quite unexpectedly around the turn of the century, during the golden days of research into radioactivity. Sensitive instruments recorded radiation even when there was no radium or uranium nearby, and it was soon found that this new radiation did not originate on Earth but came from somewhere outside. The pursuit of this cosmic radiation has led us first to high mountains, then to the stratosphere, and finally to great distances in space probes so that the locally produced effects could be eliminated and our attention focused on the primary rays themselves.

For about twenty years cosmic ray research was the only arena for the discovery of new subatomic particles. As we now know, these particles emerge from collisions between nuclei traveling at more than half the speed of light. Some of the collision energy reappears as the mass of the new particles, following Einstein's famous formula $E = mc^2$. The first of the new particles turned up among the cosmic rays in the early 1930s, and their numbers multiplied after the war, when cosmic ray research expanded rapidly. The way to the modern era of elementary particle physics was paved during those years, and cosmic rays were effectively displaced as the source of new particles only with the advent of the giant accelerators in the 1950s. Yet even the most energetic particles from these accelerators are still many millions of times less energetic than some that can be found (but rarely and uncontrollably) among the cosmic rays.

At the same time as our knowledge of the primary cosmic rays was growing in the years after 1950, radio astronomy was becoming firmly established. With observations from above the atmosphere, the scope of astronomy broadened to include the infrared, ultraviolet, and X-ray regions of the spectrum. It has been found that much of this radiation emerges from processes that also produce fast particles, and cosmic rays are now an important component of contemporary astrophysics.

Although the main focus of research on cosmic rays is now their nature and origin, we have also discovered that these high-speed particles leave clear signatures on Earth as they penetrate into the atmosphere and collide with atoms. Nuclear debris from cosmic ray collisions can be traced from the atmosphere through some unexpected but fascinating paths. For example, radiocarbon dating, so important in archaeology, owes its existence to the cosmic rays. I will explore some of these topics later in the book, but first I will retrace the historical path through various discoveries, some the result of design but others quite accidental, that led to the detection of cosmic rays.

The late nineteenth century was a time of consolidation in physics. Gravity was thought to be well understood, and Maxwell's new theory of electromagnetism tied together many previously separate topics. Although some puzzles had not been solved, there was a general feeling that the major discoveries had been made and all that remained was to pursue various quantities to their next decimal places.

Among the topics of research was the electrical conductivity of gases, which was thought to be interesting but scarcely revolutionary. As so often happens in science, however, this study began to follow unpredictable directions. The incandescent lamp provided the first of the surprises. This device contains a small, tightly coiled metal filament that emits light when an electric current flows through it. The filament also boils off electrons, but these negatively charged particles soon return to the filament, which would otherwise build up a positive charge. If, though, the glass container enclosing the filament has a metal probe inside, and if a high-voltage battery is connected at its negative terminal to the filament and its positive terminal to the probe, electrons from the filament will travel directly through the gas to the probe. In this arrangement, the filament or *cathode* and the other probe (or electrode), the *anode,* are the major components of the *cathode-ray tube* (or CRT), used in every television set.

The CRT, developed in the 1880s, became a major tool for studying electrical conduction through gases because the stream of cathode rays could be deflected by electric and magnetic fields. The nature of these rays was revealed by J. J. Thomson in the Cavendish Laboratory at Cambridge University. Until that time, as Thomson put it, "the most diverse opinions [were] held as to these rays," but his 1897 paper described quantitatively how cathode rays—now known as electrons—behave like particles. Thomson's discovery was a turning point in modern physics.

As electrons travel through the gas in the CRT, as shown in Figure 1.1, they knock other electrons off some of the gas atoms, which then become positively charged *ions.* In their turn, these ions are accelerated by the high voltage and constitute the *positive rays* also studied by Thomson. From this research came the discovery of *isotopes,* atoms of the same element with the same chemical properties (in terms of how they combine with other atoms to form molecules) but different masses.

From the discovery of radioactivity through the invention of the cyclotron in the 1930s to the modern era of nuclear research, a central thread has been the development of ways to detect and identify high-speed particles. Because most cosmic rays are electrons and nuclei and travel with speeds close to the speed of light,

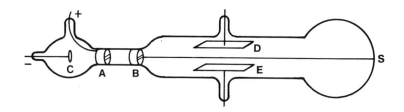

Figure 1.1   *J. J. Thomson's cathode ray tube. The negative terminal of the high voltage supply is connected to the cathode (C), and the positive terminal to the anode (A). Accelerated electrons pass through slits in the anode and another screen (B) and travel until they strike the end of the tube (S), which is coated with a fluorescent material. The cathode ray beam can be deflected by application of a suitable voltage across the additional plates (D and E) or by using a magnet.*

their detection and speed measurement draw on methods devised for nuclear physics experiments at giant accelerators.

The turn of the century was a particularly exciting time for uncovering the intricacies of atomic structure, the properties of electrons, and the different types of ions. Even though Maxwell's new theory of electromagnetism seemed to provide many of the answers, the nature of light and its emission continued to be the subjects of research. In the 1890s, Wilhelm Conrad Roentgen at the University of Wurtzburg was studying the phenomena of phosphorescence and fluorescence, in which materials emit light on their own. When using a cathode ray tube, Roentgen noted that the glass fluoresced where the cathode rays hit. Fluorescence was also observed near the cathode ray tube when a glass plate (or screen) with a sensitized coating was used. One day in November 1895, Roentgen found that his fluorescent plate glowed even when at a considerable distance from the cathode ray tube and even when the tube was completely covered by black cardboard. Clearly, some radiation was reaching his plate and penetrating the paper and other materials that covered the plate. Inserting his hand in the way, he could see the outline of his bones on the fluorescent screen. Roentgen had discovered X-rays, and in 1902 he received the first Nobel Prize for physics.

Worldwide interest was aroused. Henri Becquerel, who heard Roentgen's first presentation at the Académie des Sciences, was prompted to resume his own work on fluorescence. Less than a month later, while using compounds of uranium, Becquerel noticed that photographic materials were fogged, even when fully wrapped to protect them against fluorescence from the chemicals he was testing or from sunlight. It soon became clear that the uranium itself was emitting some sort of radiation without the stimulation of sunlight. Over the next few years this property of radioactivity was found in thorium, and then, by Pierre and Marie Sklodowska Curie, in previously unidentified chemical elements that they named radium and polonium.

X-rays and radioactive emanations ionize gases, enabling the gas to conduct electricity. This effect provides the means for detecting and measuring radiation, and electroscopes and electrometers came into wide use in those early days when X-rays and radioactive materials were being explored with such excitement. Two simple types of electroscope, shown in Figure 1.2, measure the intensity of radiation through its effect on a stored electric charge. As radioactive elements came to be identified, it was found that each element displayed a radioactive power that decreased over time. Each substance took a different amount of time for its activity to decrease to half of the starting value. This *half-life*, an inherent

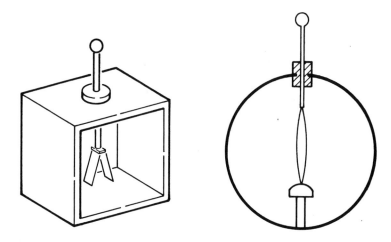

Figure 1.2 *Electroscopes: the type on the left, with two gold leaves, is still often used for lecture demonstrations; in the Wulf electroscope (right), two wires are used instead of the gold leaves. When an electroscope is given an electric charge, the leaves or wires repel each other and stand apart. Radiation can ionize some of the air in the electroscope and allow the charge to leak away, as shown by the wires or leaves slowly coming back together.*

property of each radioactive element or isotope, is unaffected by pressure, temperature, or chemical combination.

As researchers puzzled over the newly discovered radiations, they encountered a troubling experimental fact. No matter how well they made their electroscopes and no matter how good the insulation, the electric charge continued to leak away, even when there was no obvious nearby source of X-rays or radioactivity. More strenuous efforts reduced the rate of leakage but could not completely eliminate it—not even when the electroscope was surrounded by more than two inches of lead. Whatever was responsible for this leakage had a remarkably penetrating power.

There followed numerous attempts to measure the rate of residual leakage in different locations. To reduce the effect of sources of radiation in the ground, for example, electroscopes were carried to the tops of tall buildings. The inconclusive results were attributed to radioactivity from building materials. In 1910 Father Thomas Wulf took his electroscope about 900 feet up the Eiffel Tower and observed a reduction of 64 percent in the leakage rate. Wulf had expected that radiation from the ground would have decreased by far more than this because of absorption in the air, and he deduced that radiation from the ground (gradually decreasing with altitude) was probably competing with radiation coming down through the atmosphere.

The obvious next step was to go to greater heights, and several investigators did just this, using balloons. An Austrian physicist, Victor Hess, starting in 1911, was the first to produce decisive results from balloon flights in which he ascended along with his electroscopes (Figure 1.3) Hess noted that the intensity of the ionizing radiation first decreased as the balloons went up, but by about 5,000 feet the radiation was definitely more intense than at sea level. By the maximum height of 17,500 feet, the radiation had increased several times. Hess asserted that the explanation was "an extra-terrestrial source of penetrating radiation."

Further flights, notably by Werner Kolhörster of the Physikalisch-Technische Reichsanstalt in Berlin and later by the Americans Robert Millikan and Ira S. Bowen, confirmed the findings. By now the balloon altitudes were great enough that the electrometers needed to be operated by remote control and the measurements automatically recorded. After his first measurements, however, Millikan concluded that "there exists no radiation of cosmic origin" because even during balloon flights over San Antonio, Texas, the radiation seemed to be "not more than 25% of that found by European observers" (1926, p. 360). We now know that Millikan's findings of low intensities, though correct, were produced by the Earth's magnetic field, which is very different over Texas and Europe. Millikan, a man who made major contributions to physics, was given to strongly held views, sometimes erroneous. In this case, he reversed his judgment after making further high-altitude measurements. By 1926 Millikan finally came to agree that the radiation did not originate on the Earth, and he introduced the name *cosmic rays*.

From these early observations to the sophisticated and wide-ranging research of today stretches a tangled trail of puzzles, solutions, and new puzzles, pursued with great technical ingenuity and frequent shifts in emphasis. In later chapters I review the current state of each of the main areas of cosmic ray research; here I present a brief overview that relates the different areas.

It has now been firmly established that cosmic rays fall into two broad categories. One class, in the minority, comes from the sun, as shown by the relatively minor changes in cosmic ray flux between day and night. Increases in cosmic ray intensity are detected in conjunction with large solar flares—giant eruptions that can be seen (by telescope) on the face of the sun. But most cosmic rays come from much farther away. These galactic cosmic rays are now thought to originate in our own (Milky Way) galaxy, the region that contains about one hundred billion stars, including our sun and solar system. The entire galaxy is about 100,000 light years in diameter, and we are located approximately on its equatorial plane,

Figure 1.3
Victor Hess, discoverer of the cosmic rays, after his 1912 balloon flight that reached an altitude of 17,500 feet. (Photograph courtesy of Martin A. Pomerantz.)

about halfway from the center. Stars and dust are mostly concentrated in the equatorial region of the galaxy. We call this region the Milky Way, and we can see it extending across the sky when we look from places that are free of air pollution and bright city lights.

Within the galaxy, the sources of the cosmic rays are still not well identified. Many, perhaps most, come from unusual stars such as supernovas, which have exploded and hurled their debris into the interstellar space. Just how and where the rapid acceleration of the particles takes place is not settled. Yet many measurements of radiation from supernova remnants clearly display the presence of high-speed electrons. The best known of these supernova remnants (SNR) is the Crab nebula in the constellation Taurus. So well has this object been studied that, as the theorist Geoffrey Burbidge put it, astrophysics at one time consisted of the Crab nebula and everything else. The explosion of the Crab was well documented in 1054 A.D. by astronomers in China, Korea, and Japan. It was a spectacular event. The Crab was brighter than any star in the night sky—comparable to Venus, as the Chinese noted—so bright that it could be seen in daylight for twenty-three days. By now the Crab has been observed in all parts of the electromagnetic spectrum, from radio to gamma rays.

Understanding events like the Crab outburst requires a familiarity with the nuclear physics of fusion processes that take place deep within stars, providing the energy that keeps the stars glowing. This knowledge, in turn, depends on laboratory studies of nuclei and their methods of interaction. Nuclear astrophysics has become a major research area, and two of its pioneers, Subramanyan Chandrasekhar of the University of Chicago and Willy Fowler of Cal Tech, shared the Nobel Prize for physics in 1984.

Most of the galactic cosmic ray particles originate in our galaxy, but growing evidence indicates that some of the highest-energy particles come to us from outside—presumably originating in the myriad distant galaxies that telescopes have revealed. As cosmic rays approach the Earth, they encounter its magnetic field. The Earth's behavior as a giant magnet has been known for centuries; it provides the basis for simple magnetic compasses. The paths of electrically charged particles are deflected in a magnetic field, and near the Earth they become highly convoluted, depending in detail on each particle's speed and arrival direction. As a result, the numbers and speeds of cosmic rays close to Earth will not be the same as those measured far away. In order to understand the properties of the *primary* cosmic rays, we must make corrections to our local measurements. Between the 1930s and the 1960s, many cosmic ray scientists directed their attention to these geophysical effects.

As cosmic ray studies progressed, it was found that the energy of individual particles, measured in electron volts (eV), was often very high. At room temperature the typical energy of an air molecule is a few hundredths of an electron volt, whereas cosmic rays have energies well above one billion (10^9) electron volts, and the highest energy yet measured is around 10^{20} eV. The processes by which such truly astronomical energies are produced have not been identified with any certainty, although theorists have suggested several possible mechanisms.

When high-energy cosmic ray particles enter the Earth's upper atmosphere, they collide with atoms of the air. Heavy cosmic ray particles, such as nuclei of helium, carbon, or other atoms, are generally fragmented in these collisions, and particles can lose significant fractions of their energy. Emerging from these collisions are streams of particles; some are nuclear fragments and others are unfamiliar particles called *mesons*. These are not constituents of the matter that we normally encounter; they are radioactive and their half-lives are short, usually less than one-millionth of a second. When they decay or collide with air atoms, more particles including electrons are produced. The electrons can radiate photons, which produce yet more electrons so that cascades develop as this process is repeated over many generations down through the atmosphere, until showers of billions of particles (mostly electrons) can be detected simultaneously at ground level. These *extensive air showers* provide the only means of detecting cosmic ray particles of the highest energies.

There was a period of intense activity when cosmic ray collisions were studied for their own interest. Because the colliding particles were far more energetic than any that could be produced by cyclotrons, the highest-energy nuclear collisions could be studied only in this way. Many new short-lived particles were discovered during this period, which lasted from the 1930s until the start-up of the Cosmotron at Brookhaven National Laboratory in 1953 and the Bevatron at the University of California at Berkeley in 1955. Since then particle physics has become essentially a laboratory science, with the exception of the highest-energy particles, which are still the sole property of cosmic ray researchers.

In deciphering the physics of high-energy collisions and fundamental particles, theorists have realized that these particles must have played a critical role during the earliest stages of the explosive "big bang" from which our universe is thought to have started, about 15 to 20 billion years ago. In the tiniest fraction of a second, particles and radiation must have interacted in that hot and expanding turmoil. Calculations of the subsequent stages can be tested against present-day observations. The result of one such calculation is the amount of helium produced from the fusion of hydrogen in the early big bang. Helium is continually being produced in nuclear reactions in stars, but not in sufficient quantities to account for all that we observe; production in the big bang seems to solve this problem. The calculation of helium production is very sensitive to assumptions regarding the properties of some elementary particles. Although present-day cosmic ray particles are not directly involved in such calculations, there is no doubt that cosmic ray research paved the way for the modern study of elementary particles.

In a later chapter I will recall this fascinating period in cosmic ray history, but first a more basic question must be addressed: How do we detect and identify cosmic rays, and what is the astrophysical picture we can draw from our findings?

Chapter 2

Identifying Cosmic Rays

*A*fter the accidental discovery of the leakage of electric charge from electroscopes, it took only a few years before researchers generally agreed that this leakage was indeed caused by a penetrating radiation, of cosmic origin, quite different from X-rays and radioactivity. Radioactive atoms had been found to produce three different types of radiation, named alpha, beta, and gamma by Ernest Rutherford in 1903. The initial classification was based on the degree to which the radiation penetrated lead (which was widely used for shielding) and on the effects seen when magnets were brought near. Rutherford, among others, showed that alpha and beta radiations consisted of particles; alpha particles are identical to the nuclei of helium atoms, and beta particles are the same as electrons and cathode rays. Gamma rays seemed to be more like penetrating X-rays than particles. The newly discovered cosmic rays, however, appeared to have properties that differed from any of these. It was many years before the nature of the cosmic rays (CR) came to be understood, because there are so many different types and their relative proportions change in a complex way as they penetrate the atmosphere.

The primary cosmic rays, both solar CR and galactic CR, that approach the Earth from outside the atmosphere collide with atoms in the stratosphere and produce secondary CR which may, in turn, have their own collisions. Such secondary CR were responsible for the electroscope leakage. Most primary CR are protons, the nuclei of hydrogen atoms, which is not surprising if one considers that hydrogen is by far the most abundant element in the universe. On Earth, there is little hydrogen in the atmosphere but a lot in the molecules of ocean water; hydrogen constitutes about 94 percent of the atoms in the sun and is the element from which other (heavier) atoms form during fusion reactions at high temperatures.

Some fusions occurred very early in the development of the universe, in the expansion after the big bang, but others are still taking place, in the interiors of stars and during stellar explosions. The formation of complex nuclei in fusion reactions is termed *nucleosynthesis*.

After hydrogen, helium is the next most abundant element in the sun and stars, and also in cosmic radiation. Among cosmic rays, helium is roughly ten times less abundant than hydrogen. The CR helium nuclei are often called alpha particles, continuing the older nomenclature. In the solar system, all atoms other than hydrogen and helium together amount to only about 0.1 percent of the hydrogen abundance. From this variety of atoms we ourselves are made; indeed, our very existence is excellent testimony to the process of nucleosynthesis.

The primary CR beam also contains small numbers of electrons (less than 2 percent) and gamma rays. Classification of the gamma rays with the "particles" is perhaps a matter of taste, for gamma rays are the highest-energy form of electromagnetic (e-m) radiation. The spectrum of the e-m radiation extends smoothly from the longest-wavelength radio waves to the shortest-wavelength gamma rays (the intermediate visible wavelengths provide the observational base for classical astronomy). The energy in e-m waves frequently behaves as though it is concentrated in bundles (or *photons*) rather than being spread out, and the photons behave in many ways like particles. Historically, the debate over the nature of light (wave or particle?) has drifted back and forth. Today we use both descriptions, depending on the phenomenon being described.

The main difference between "particle" cosmic rays and gamma rays is that gamma rays have neither mass nor electric charge. Protons, alpha particles, and heavier nuclei have positive charges, and the electrons have a negative charge. Gamma ray photons are not the only zero-mass CR particles; *neutrinos*, which are released in many nuclear reactions such as those taking place in the sun, are also produced in some radioactive decays. Having neither mass nor electric charge, they continually flood the Earth, passing through us at all times. Their detection is so difficult that although their existence was first hypothesized in the early 1930s, they were not discovered until 1953.

Collisions of nuclear cosmic rays in the atmosphere produce *mesons* (mostly *pions*, previously termed π-mesons), which in turn produce *muons* (originally termed μ-mesons). Charged pions (positive and negative) turn into muons and neutrinos in much less than one-millionth of a second. Not quite as rapidly, muons turn into electrons and more neutrinos. Neutral (uncharged) pions turn—in less than 10^{-16} of a second—into two gamma rays, which in turn produce electrons, which in the atmosphere produce more gamma rays, and so forth. This ever-changing cascade of electrons, neutrinos, and gamma rays, intertwined with muons, moves through the atmosphere at almost the speed of light. Researchers were not able to interpret sea-level and low-altitude CR observations correctly until they were able to carry out measurements at very high altitudes, where the primary CR alone could be detected. This development, together with the discovery of pions during the period 1946–1950, finally enabled researchers to understand the nature of secondary CR effects at different altitudes.

It is easy today to sit down and write about all these developments, but many of the early results were confusing and clarity came slowly. The exploratory phase lasted for nearly fifty years. During that time experimental tools slowly improved, and as a result the complexity of the processes was gradually recognized. Even a brief description of the experimental methods shows how varied and ingenious were the approaches and devices. It is no mean trick to measure the mass, electric charge, and speed of a particle that weighs less than $1/100,000,000,000,000,000,000,000$ (10^{-23}) of a gram and zips through apparatus in less than 10^{-8} of a second. Our handle on each particle is its electric charge, but this, too, is tiny. It takes a stream of 10^{19} electrons per second to make up a current of one amp (a typical electric toaster uses 10 amps). The arrival of one CR particle each second constitutes a current of around 10^{-19} amps. How can such incredibly small particles be detected and identified?

Charged particles can be detected through the ionization they produce. As noted earlier, charged particles collide with the nuclei and electrons in the materials they penetrate and knock some electrons off their parent atoms. These electrons can be collected and constitute a minute and measurable electric current. Some electrons are fast enough to produce secondary electrons or other detectable effects. And some CR particles may collide with atomic nuclei. The CR particle might then be fragmented, or the target nucleus may have bits thrown out that will also ionize. These nuclear collisions have been studied for their own interest (revealing the properties of the colliding articles), and they can also be used to measure the energies of the article projectiles.

A particle's ionization depends on its electric charge (Z) and speed (v). Ionization depends mainly on the ratio Z^2/v^2, which shows that highly charged particles ionize much more than lighter, less heavily charged particles. For instance, an alpha particle (with $Z = 2$) produces four times the ionization of a proton ($Z = 1$) with the same speed, and a carbon nucleus (having $Z = 6$) produces thirty-six times more ionization. This makes it easier to detect heavier particles, even though they are much rarer. Speed dependence indicates that slow particles ionize much more heavily than

Figure 2.1
The ionization chamber used by A. H. Compton for his wordwide surveys. The central chamber was heavily shielded by layers of bronze and lead. (From **Physical Review 43** *[1933]: 390.)*

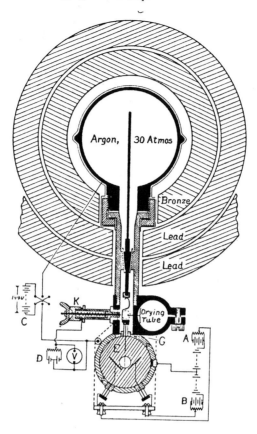

do fast particles: the slower the particle, the more time it spends near each atom in detectors, thus allowing it to lose more of the energy by which it can be identified.

CR particles have *kinetic* energy (energy resulting from motion), which is the source of the energy needed for ionization. Ionization continues as long as the particle has kinetic energy to lose. Ultimately, a particle slows down, loses all its energy, and comes to rest. The stopping distance is the *range,* a measure of the original kinetic energy. Thus if we can measure the range of a particle, we can infer its initial energy by using calibrated range-energy tables. Kinetic energy depends on both the mass of the particle and its speed. If we can determine the energy (by measuring the range) and the speed (from the ionization), then we can identify the particle; that is, we can figure out its electric charge and mass. Other identification methods have to be used for particles that are too fast to stop. In general, to identify a particle we need two different measurements that depend in different ways on mass, charge, and speed.

There are two broad classes of particle detectors. In one, the actual path (or track) of the particle is made visible, and the range can be found directly. In addition, ionization can be measured along the track and changes detected as the particle slows down. In the other type of detector, individual tracks are not seen but the particles' ionization can be measured, either directly or indirectly; the path of each particle is then determined by noting its passage through several detectors in sequence.

The discovery of cosmic rays was based on ionization in an electroscope. Pioneering experiments, carried out by Millikan before 1920 and by Arthur Compton during the 1930s, also used ionization chambers. In all these studies, the total ionization in a closed container was monitored. This method records the total flow of cosmic rays through the chamber, and thus is a simple way of surveying the variations of CR intensity with altitude (as was done by Millikan) or variations around the Earth, especially at different latitudes (as measured by Compton).

Compton's ionization chamber, shown in Figure 2.1, had a 1-foot diameter and was shielded against local radioactivity by layers of lead. In the center, a spherical container filled with the rare gas argon held a probe connected to a high voltage source. Electrical leakage between the probe and the container's body was monitored to detect ionization produced by cosmic rays passing through the argon. Although such ionization chambers have largely been replaced by other devices, they played a key role in early discoveries. They were soon supplemented by track detectors and later by complicated electronic systems.

The earliest electrical detector for single particles was devised around 1911 by Hans Geiger, an assistant of Rutherford. In 1929 Geiger and his student Walther Müller introduced an improved version. The design was simple: a thin wire is fixed along the central axis of a metal cylinder but is insulated from direct contact with the cylinder. The cylinder is filled with a low-pressure gas, often argon. A high-voltage battery is then connected between the wire and the cylinder. Because of the gas's low conductivity, only a minute current normally flows around the circuit. But penetration by a charged particle produces ionization in the gas; ions and electrons are accelerated to the wire and the cylinder by the high voltage. Along the way, more ions are produced by electron collisions. An avalanche of ions and electrons arriving at the wire and cylinder constitutes a brief but measurable electrical current, and in this way the passage of the particle can be noted.

Developed at about the same time as the Geiger counter, the Wilson cloud chamber was for many years the most widely used track detector. C. T. R. Wilson, a student of J. J. Thomson at Cambridge and one of the discoverers of electroscope leakage, was studying the condensation of water vapor. He realized that rapid expansion of gas would produce cooling, as when air escapes from a tire. Vapor condenses when it cools sufficiently, and Wilson found that the vapor condensed on ions and small dust particles. By 1911 he had devised an "expansion chamber" (now usually called the Wilson cloud chamber) to study condensation under controlled conditions of expansion and cooling. He showed that X-rays and radioactive particles produced many ions that served as centers for vapor condensation, forming tracks of water droplets that could easily be seen.

Cloud chambers were used extensively in CR and nuclear physics, and major improvements in technique were implemented during the 1930s and in the subsequent twenty years. One of the most important findings with this tool came in 1932, when Carl Anderson and Seth Neddermeyer of Caltech discovered the *positron,* the first example of a particle of antimatter (Figure 2.2). The

Figure 2.2   *Cloud chamber track of the positron, discovered by Carl Anderson. Use of a strong magnetic field caused charged particles to follow curved paths. The positron entered the chamber at the lower left and traveled up through the lead plate across the middle of the chamber. In traversing the lead the positron lost energy, so its track was more strongly curved when it traveled farther up. Curvature measurements showed that the particle had an energy of 63 MeV before entering the lead and 23 MeV afterwards. (From* **Physical Review** *43 [1933]: 491, courtesy of the Archives, California Institute of Technology.)*

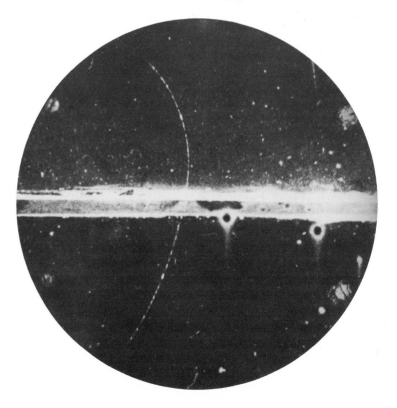

possible existence of such particles had been predicted by P. A. M. Dirac in 1929, in his abstract quantum theory calculations. A few years later, Anderson and Neddermeyer also found the first particle whose mass was intermediate between those of the electron and the proton. (These particles were first called mesotrons, then mesons, and are now known as muons.) Anderson shared the 1936 Nobel Prize with Victor Hess, belatedly honored for his discovery of cosmic rays. Dirac, along with Erwin Schrödinger and Werner Heisenberg (co-inventors of quantum mechanics), received the 1932 award.

Because tracks in cloud chambers dissipate rapidly, the only good tracks are those that form during a brief interval around the expansion. In many early experiments cloud chambers were triggered randomly; it was a lucky accident when a cosmic ray happened to pass through the chamber during the sensitive time. An innovation by Walter Bothe changed this situation. Starting in 1924 Bothe and Geiger used two Geiger counters for the simultaneous detection of electrons and X-rays in scattering experiments, and Werner Kohlhöster and Bothe then adapted the method to detect cosmic rays. Bruno Rossi, in Florence, greatly increased the sensitivity of this method by developing a new electrical circuit to analyze the signals from a CR telescope—two or three Geiger counters set in a row (Figure 2.3). If a single cosmic ray went directly through the counters, each would give a signal and a *coincidence* would be registered. If the counters were triggered by separate particles, the electrical signals would not in general be close together.

Figure 2.3
The Geiger-Müller telescope. A coincidence is registered only when a single particle (A) passes through both counters. It is unlikely that separate particles (B and C) will go through the counters within a very short interval.

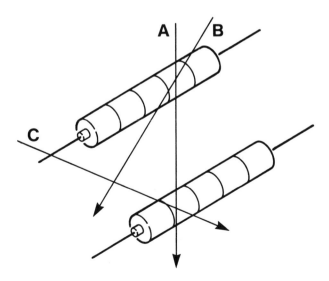

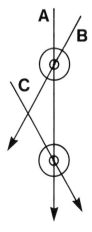

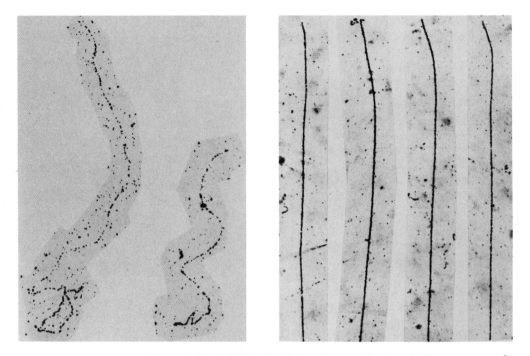

Figure 2.4   *(left) Tracks of slow electrons showing the large amounts of multiple scattering that result from the electrons' small mass. Note also the increase in ionization as the electrons slow down. The tracks shown are about 0.2 mm long. (right) Proton tracks (about 0.2 mm long) generally show a much higher degree of ionization and are much straighter, even to the very end where they stop (at top of photo). (Photograph courtesy of Peter Fowler, University of Bristol.)*

G. P. S. Occhialini had worked with Rossi in Italy, and when he joined P. M. S. Blackett at the University of Manchester, he applied the coincidence method to Blackett's cloud chamber. Geiger–Müller counters were placed directly above and below the chamber, which was expanded and photographed only if a coincidence was recorded between the counters.

Predating both Geiger counters and cloud chambers were photographic materials, which provided the earliest track detectors. Photography was invented in the mid-nineteenth century, long before cosmic rays were discovered or the nature of atoms and electrons was known. When light falls on a photographic emulsion (a thin layer of gelatin containing crystals of silver bromide), it produces submicroscopic changes that show up after suitable chemical treatment (development). Tiny silver grains are produced where the light was absorbed. When fast charged particles pass through a photographic emulsion, they produce similar changes, and the track of the particle will show up (after development) as a trail of silver grains (Figure 2.4). "Fast" in this context really does mean fast; an alpha particle whose speed is less than about 1 percent of the speed of light will produce a track consisting of only one grain.

Heavy particles with large electric charges leave thick tracks, made up of thousands of grains. Lighter particles with smaller charges produce thin tracks, often with the grains well separated. The particle's speed also affects the shape of the track. Light and slow particles are buffeted around, or scattered, by collisions with the atoms in the emulsion, because of electric forces between each particle and the nuclei in the emulsion atoms; the resulting tracks are full of wiggles. Fast and heavy particles, having greater momentum (mass times speed) leave much straighter tracks (for the same reason that a fast and heavy football player is not easy to deflect). The "straightness" of a track can be put into quantitative terms and reveals the momentum of the particle. Microscopes with a typical magnification of $1,000\times$ are used for making track measurements, which is done either by counting grains or by measuring the scattering along a track. The grains one can see in the photomicrographs are around $1/1,000$ mm (about $1/25,000$ of an inch) in diameter.

Photographic plates, measuring a few inches on a side, were used by Ernest Rutherford in his early experiments on alpha particles. With these plates he could count alpha particles, but the emulsions were too thin for long tracks to be seen. The longest-range track of an alpha particle from a radioactive nucleus is about 50 microns ($1/20$ mm, or $1/500$ of an inch). Great improvements in emulsion sensitivity were achieved after World War II.

One application of the photographic method, largely in the biological sciences, has been *autoradiography*. Tissue samples with radioactive isotope tracers are either held against coated films or have liquid photographic emulsion poured over them. Development, after a period of time that depends on the isotope half-life, shows tracks from the tracers that reveal where in the tissue the isotopes have been transported by biological processes.

The early use of photographic plates in balloon flights to study cosmic rays set a pattern that was followed for many years. In November 1935, Captain Albert W. Stevens and Captain Orvil J. Anderson of the U.S. Army carried special Eastman Kodak plates into the stratosphere aboard Explorer II when they set the altitude record for manned balloon flights at 72,395 feet (Figure 2.5). Those photographic plates, developed and examined by T. R. Wilkins of the University of Rochester, clearly showed CR tracks, probably the first record of primary CR tracks (Figure 2.6).

Two Viennese scientists, Marietta Blau and Herta Wam-

Figure 2.5   *Explorer II. Launched from Rapid City, South Dakota, on November 11, 1935, it touched down near White Lake just over 8 hours later, after a record high-altitude flight. Cosmic ray detecting equipment carried on board included ionization chambers and Geiger counters for W. F. G. Swan of the Bartol Research Foundation, and photographic plates for T. R. Wilkins of the University of Rochester. (Photograph courtesy of Richard H. Stewart and Captain James Haizlip, © 1936 National Geographic Society.)*

would have to include all of astronomy. Where, along this seamless spectrum, should the line be drawn? For the purposes of this book, a convenient division can be made to include gamma rays with energies above half a million electron volts, because these high-energy photons can be produced by cosmic rays.

The detection of gamma rays draws on many of the techniques developed for particle detection. Gamma rays can produce ionization, and different types of ionization detectors can be used depending on the gamma ray energy. High-energy gamma rays (above about a million electron volts) can convert into *pairs* of electrons and positrons. These particles, in turn, can be detected in ionization detectors or produce tracks in a *spark chamber*. In this device, thin parallel metallic sheets are connected to high positive and negative voltages. When a fast particle passes through the sheets, it creates ionization that causes a spark to jump between adjacent sheets, showing where the particle has traveled; successive sparks mark out the particle's track. As with cloud chambers, a spark chamber must be photographed from two angles simultaneously, to generate stereoscopic views from which a three-dimensional picture of the sparks can be reconstructed by using a computer. The midline between tracks of an electron and a positron in a pair indicates the direction of the original gamma ray. Like starlight, gamma rays travel along essentially straight paths, so each arrival direction is unaffected by the Earth's magnetic field. The angle between the two tracks is a measure of gamma ray energy; smaller angles correspond to higher energies.

Extremely high-energy gamma rays (above about 10^{15} eV) can produce cascades of electrons, positrons, and secondary gamma rays in the atmosphere. By the time this cascade reaches ground level, a million or more particles arrive almost simultaneously and spread out over a large area. These extensive air showers are the only means of detecting the very highest-energy gamma rays and also charged particles, which produce similar showers. I explore them in greater detail in Chapters 6 and 9; for now the important point is that the sea-level detection of showers uses ionization and scintillation counters similar to those described earlier.

The detection of cosmic rays, as we have seen, has required the invention of a whole arsenal of new techniques. Over the years the detectors have grown from small photographic plates and electroscopes to giant electronic systems weighing tons—but still capable of being carried into orbit on satellites. In the following chapters I look at the results obtained from this exploration and the inferences that can be drawn about the nature and origin of cosmic rays.

The Earth's Magnetic Influence

*A*part from airplane flights, we live our lives on the Earth's surface, drawing comfort from its seeming solidity. It should be no surprise, therefore, that the earliest thinkers considered the Earth the center of the universe, with the sun, planets, and stars traveling sedately around us in their divinely assigned orbits. Only relatively recently, on the scale of human history, did we recognize that we, too, are moving, with the Earth circling the sun. The entire solar system orbits around the center of our Milky Way galaxy, and, on an even larger scale, the galaxy participates in the general expansion of the universe.

To reach this understanding we have relied on astronomical observations, made first with our unaided eyes and then with telescopes. In all these observations we have depended on an important fact: with minor exceptions, the light we see (such as that from the stars) has come to us along straight-line paths. That is, we generally do not look in one direction to see light coming from other directions. With cosmic rays, however, the situation is almost completely reversed.

Observing cosmic rays from the Earth's surface is like making astronomical observations from the bottom of an ocean. In such a situation, the constant swirling of the ocean surface would distort all images and make them quiver. In addition, the observer would have to correct for the bending (refraction) of light as it passed from the air into the ocean to determine the true direction of an astronomical object. An occasional oil spill would change the amount of light absorbed, as would changes in the quantity of plankton, and the degree of this absorption would vary with the light's wavelength (color). The resulting pictures constructed of the heavens would certainly be distorted.

Cosmic rays detected in apparatus at the Earth's surface or even in the stratosphere are either survivors from primary CR or offspring of those survivors. They have survived collisions with gas atoms in interstellar or interplanetary space, and they also may have collided in the atmosphere. Raw observations thus provide only the first stage in constructing a picture of the CR beam and its journey from distant sources to the Earth.

Just as important as the effects of collisions is the influence of the Earth's magnetic field. In many ways the Earth behaves as though it had a strong bar magnet buried deep within it, about 200 km from its center and not quite aligned with the axis of the Earth's rotation. The extensions to this imaginary magnet come to the Earth's surface at one geomagnetic pole near 79°N, 69°W, north of Thule in Greenland, and the other close to 79°S, 110°E, in the Australian Antarctic. Worldwide surveys of this field have led to the mapping of a grid of imaginary lines of magnetic latitude and longitude, similar to the geographic parallels of latitude and meridians of longitude.

What is meant by a magnetic field? The term "field" denotes a region of space where a force (in this case, magnetic) can be detected and where the strength of that force can be specified at all points. The field strength may be the same throughout a region or it may vary greatly, depending on the nature of its source. It is not necessary for anything visible or tangible to support the field: the force of a simple magnet can be *felt* when we try to pull an iron object away from it.

Electromagnetic (e-m) waves carry programs from transmitters to radio and television antennas. All e-m waves travel with the speed of light, carried by electric and magnetic fields whose strength oscillates very rapidly. If we could momentarily freeze a wave, we would find that it forms an endlessly repeating pattern along which places of maximum field strength are regularly spaced one *wavelength* apart. The fields oscillate many times each second with a well-defined *frequency*, determined by a general relation between the wavelength, the frequency, and the speed of light. In television transmission the fields change with frequencies of around 100 million oscillations per second (100 million hertz or 1 MHz), and the wavelengths are around 300 cm. FM radio uses similar frequencies, but AM radio frequencies are lower, around 600,000 hertz (600 kHz). The visible part of the e-m spectrum has wavelengths extending from about 4×10^{-5} cm at the blue end to 7×10^{-5} cm in the deep red. Much shorter wavelengths include ultraviolet, X-, and gamma rays, while the longer wavelengths extend into the infrared and then radio.

The Earth's magnetic field does not oscillate rapidly as an e-m wave does but changes slowly over tens of thousands of years. It extends far into space, with steadily decreasing strength; the field drops by a factor of two at a distance of a thousand kilometers out from the Earth's surface. Farther out the field continues to decrease until it encounters the ionized gas that is continually spreading out from the sun. This ionized gas, called the *solar wind,* was discovered as recently as the early 1960s, although its presence had been suspected many years earlier.

The interaction between the solar wind and the Earth's magnetic field is now studied with satellites and long-range space probes, but a starting point is an understanding of the basic phenomenon of magnetism. The behavior of magnetic rocks called lodestones has been known since ancient times. These have been used for navigation since at least the twelfth century, and simple magnetic compasses, such as the kind one attaches to a dashboard, are still in use. Magnetic compasses do not point accurately to true north, however; one must compensate for local effects as well as for the displacement of the north magnetic pole from the geographic pole.

The first systematic study of magnetism was published in 1600 by William Gilbert, physician to Queen Elizabeth. This work had a strong influence on Johannes Kepler, who developed his theories of planetary motion under the mistaken impression that a magnetic force from the sun drove the planets in their orbits. Kepler made other incorrect assumptions that eventually canceled each other out in his calculations; as a result he did, in fact, deduce correct laws of planetary motion. But it was Newton, about sixty years later, who correctly identified gravity as the controlling force.

Edmond Halley, so well remembered for his association with the famous comet, was also renowned for his magnetic survey. He carried out this work during three voyages between 1698 and 1702 for the British Admiralty, which had a continuing interest in the improvement of navigational aids. In what was probably the first scientific expedition completely supplied and funded by a government, the Royal Navy provided Halley with a ship and all the instrumentation needed for his survey. Royal recognition came in 1701, on the command of King William to the Navy Board: "In obedience to His Majesty's Commands . . . desire and direct to be paid unto Captain Ed. Halley . . . the Sum of two Hundred Pounds . . . in consideration of his great Paines and care in the late Voyage he made for the discovering the Variation of the Needle" (*The Three Voyages of Edmond Halley,* p. 321). (The "variation" is the angle between the true north and the direction shown by the magnetic compass needle.)

The relevance of magnetism to cosmic rays is that a stream of electrically charged particles (such as CR) constitutes an electric current. The effects of a magnetic field on an electric current have been known since the pioneering work of Hans Christian Oersted and others in the early nineteenth century—research that led to the invention of electric motors and generators.

The paths followed by cosmic rays as they approach the Earth are influenced by the Earth's magnetic field. Several factors control the path of a CR: its initial direction of travel; its mass, speed, and electric charge; and how the magnetic field varies with distance from the Earth. As a result, the direction of motion changes continually as a CR approaches the Earth, resulting in a highly convoluted path. We can now see why the path of a CR particle detected by a CR "telescope" (of G-M counters, for example) is not a reliable indicator of the particle's direction before it came to Earth. Indeed, because of magnetic fields even in the great distances between planets and the greater distances between the stars, a CR path is never simply a straight line from its source to the Earth (see Figure 3.1). I elaborate on this in Chapters 5 and 7 in the discussion of galactic CR and their history.

Early surveys of the Earth's magnetic field were, of necessity, carried out close to the Earth's surface, but with the advent of the space age, direct measurements have been extended to great distances but still within the solar system. A mathematical model based on these surveys enables us to calculate in detail (with the use of computers) a CR's path through the Earth's magnetic field. The calculation is simple but laborious: start with a CR at some location far from the Earth and specify its charge, mass, speed, and direction of motion. The magnetic force that deflects the particle depends on the combination of these factors. Then, using a computer and a

Figure 3.1 *A computed cosmic ray orbit, showing the effect of the Earth's magnetic field on incoming charged particles. The complexity of each orbit depends sensitively on the particle's charge, momentum, and direction of travel. Some particles spiral around the Earth many times before descending to a low altitude. (Courtesy of Niels Lund, Danish Space Research Institute.)*

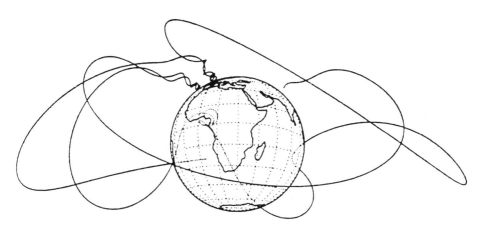

model for the field, calculate the field strength at the place and compute the force that the particle will experience because of the field. This force changes the particle's direction by a small amount, so one can calculate where the particle will be a short time later. Now repeat the calculation, allowing for the changed direction and also for any difference in field strength. Once more, calculate where the particle will be after another short interval. Then repeat the whole process—millions of times. Subtleties in the programming of such a calculation can keep the computing time (and cost) to a minimum while protecting against errors that could accumulate when a final result is based on millions of successive calculations.

In working out the solution of a theoretical physics problem (such as CR paths), you start with a few equations that relate various measurable quantities to one another. With some types of equations, however, there is no exact solution in the form of a general formula. You then have two choices: either produce numerical solutions for specific cases, or go back to the starting equations and see whether you can simplify them and then derive exact solutions. Simplified equations refer to idealized circumstances, but they can serve as a guide in working numerically with the complete equations.

The shapes of CR paths pose one such problem, with no general formula as a solution. Fortunately, it has turned out that one can solve equations for a simplified model of a magnetic field similar to the Earth's. This approach was first used by the Norwegian mathematician Carl Störmer starting in 1903, relatively soon after the electron was discovered. As he described it, his inspiration came when "my colleague Kristian Birkeland showed me some of his very beautiful and striking experiments on the movement of cathode rays in magnetic fields" (*The Polar Aurora,* preface). Störmer became interested in polar auroras, the impressive atmospheric displays in which large regions of the upper atmosphere glow with brilliant colors in changing patterns (Figure 3.2). The discovery of electrons and the strong correlation between auroral activity and solar activity (such as sunspots) stimulated speculation that auroras might be generated by electrons coming from the sun. Störmer set out what has become the classic mathematical description of electron orbits in the geomagnetic field. In his research on the aurora, he focused on relatively low-energy electrons; a later extension of the theory to high-energy particles was prompted by the growing interest in the nature and origin of cosmic rays and by measurements of CR intensity made at widely separated locations.

In 1927 and 1928, the Dutch physicist Jacob Clay published the results of his measurements of CR intensity during voyages between Genoa and the Dutch colony of Java. Using ionization

Figure 3.2 *An aurora. High-speed charged particles are guided by the magnetic field as they enter the upper atmosphere. In collisions with atmospheric atoms and molecules at typical altitudes of 180 km, energy is transferred from the incoming particles and radiated to produce the aurora. (Courtesy of the National Center for Atmospheric Research/National Science Foundation.)*

chambers, Clay found that CR intensity had its minimum value at the magnetic equator and rose by about 10 percent at northern latitudes. Meanwhile, Bothe and Kolhörster, using Geiger-Müller counters, had shown that the sea-level CR consisted of charged particles. They went on to point out that Clay's latitude effect could be understood if the *primary* CR also consisted of charged particles, and if those particles were deflected above the atmosphere by the Earth's magnetic field. Clay's results were confirmed and extended in a worldwide survey carried out by Arthur Compton of the University of Chicago between 1930 and 1933 (Figures 3.3, 3.4). Compton made measurements at sea level and also at mountain altitudes, where the latitude effect was even larger.

Figure 3.3
Victor Hess examining equipment used by Arthur Compton during his worldwide survey. Note the heavy shielding around the apparatus, housed on the ship deck. (Courtesy of Washington University Archives.)

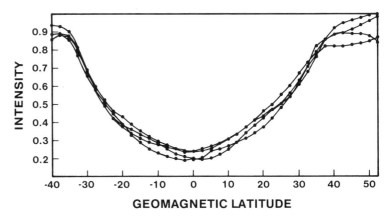

Figure 3.4
Results of Compton's survey, showing cosmic ray intensity at sea level for various geomagnetic latitudes. The four curves represent data taken during different seasons. (Adapted from Physical Review *52 [1937]: 808.)*

The theory of the geomagnetic effect on CR attracted renewed attention, first by Bruno Rossi (then at the University of Florence) in 1930, followed by the Belgian cosmologist Georges Lemaitre and the Mexican physicist Manuel Vallarta, then at MIT. Their calculations showed clearly that the latitude effect was a consequence of the fact that charged CR particles arrive randomly from all directions before they encounter the Earth's field and begin to be deflected. The calculations of Lemaitre and Vallarta and of Rossi revealed that only the fastest, most energetic particles could penetrate to the region of the magnetic equator. Somewhat less energetic particles would be detectable in the mid-latitudes, and in the polar regions there would be no restriction on the energies of the detectable primary particles. Overall, the number of particles detected should increase steadily as one moves away from the magnetic equator. The observations of Clay and Compton were thus nicely explained.

The latitude effect showed clearly that primary CR particles had to be electrically charged, but that both positive and negative charges would show the same effect. One prediction from the theory did allow the sign of the charge to be identified: each latitude has a *cutoff energy* below which no vertically arriving particles will be found. The cutoff energy is different for positive and negative particles arriving in nonvertical directions. Lower-energy particles of positive charge would penetrate if they arrived from the west, but for positive particles arriving from the east a higher energy would be needed. For negative particles, the effect would be reversed. More low-energy particles than high-energy ones are found among the cosmic rays, and thus this East-West effect predicts that, if primary CR are predominantly positively charged, more CR should be detected by a Geiger telescope directed to the west than by one directed to the east. Confirmation of the East-West effect came in 1938 from measurements carried out by Thomas H. Johnson of the Bartol Foundation and the Carnegie Institution of Washington. The inference was clear—the majority of primary CR were *positively* charged.

Improvements in the capacity and speed of computers have enabled reearchers to explore the calculation of CR paths in detail unimaginable by Störmer and even by Lemaitre and Vallarta, who worked with an early computer that used electric motors and mechanical linkages. Because a typical CR orbit can loop around the Earth many times (see Figure 3.1), it has to be followed (in the computation) through millions of small steps. An even more complex picture emerges when many CR path calculations are examined. Diagrams of many paths resemble a tangle of spaghetti. Some paths, after several loops, provide unbroken connections to the Earth's surface from the region far beyond the solar system. Some cosmic rays approach the solar system, travel along these paths, and arrive close enough to the Earth to be detected; they display the latitude and East–West effects just described. But it turns out that there is also a quite different family of paths that loop around the Earth, neither dipping down to the surface nor escaping. A particle traveling along such a path would be trapped (like a celestial Flying Dutchman) in a permanent orbit around the Earth, looping between the northern and southern hemispheres and drifting around in longitude but never escaping. Cosmic rays approaching from far away cannot connect directly to these paths, and the only way a particle can get into one of these perpetual loops is to be born in it (through some nuclear process) or be bumped into it (by a collision with another particle). For this reason, no importance had been attached to these strange orbits by CR researchers, even though their mathematical existence had been known from the calculations of Störmer, Lemaitre, and Vallarta. Then, totally unexpectedly, these orbits were rediscovered in 1958.

Under the auspices of the International Council of Scientific Unions, a coordinated international study of the Earth, the International Geophysical Year (IGY) was planned for 1957–58. Many types of observations were to be carried out simultaneously all over the world, to reach a better understanding of the Earth and its relation to the sun. The first Earth-orbit satellites were designed to extend these observations into near-Earth regions. On October 4, 1957, we awoke to the news that Sputnik I was in orbit; reports carried recordings of its high-pitched beeping signal. The political response in the United States was prompt and intense. Increased federal funds were voted for scientific research, and major efforts were undertaken to modernize the science curricula in our schools.

Sputnik I did not carry CR instruments, but a monitoring of its radio transmissions revealed properties of the ionosphere, the ionized region of the Earth's upper atmosphere. The extended gravity field of the Earth could be probed by tracking the satellite in its orbit. Sputnik II, launched in November 1957, did carry CR detectors, but nothing unusual was reported by Sergei Vernov and his colleagues at the Academy of Sciences in Moscow.

After a few initial failures the first U.S. satellites were launched in Feburary and March 1958, carrying G–M counters to monitor cosmic rays. The CR experiment, designed by James Van Allen and his group at the State University of Iowa (Figure 3.5), surprised everyone with the early results; the signals telemetered back to Earth were quite different from what had been expected. During the decade preceding the satellite voyages, rocket explorations had found that CR intensity increased up to about 24 km (as had been already known) but then decreased to a steady value at the highest altitudes reached, around 150 km. The U.S. Explorer I and

Figure 3.5   *James A. Van Allen (University of Iowa) performing preflight tests on his radiation detectors that were used in the Jet Propulsion Laboratory spacecraft Pioneer IV, launched in March 1959. These detectors produced the first high-quality data of the inner and outer radiation belts, then only recently discovered. The intensity of the primary cosmic radiation was also measured. (Photograph courtesy of J. A. Van Allen.)*

II satellites, however, were reaching far greater altitudes. In very eccentric orbits, they looped from their closest approach (perigee) at about 250 km (150 miles) above the surface out to a maximum distance (apogee) of about 2,600 km (1,600 miles) every two and a half hours. And what Van Allen noted was that cosmic rays first increased slowly with altitude, but above about 2,000 km they seemed to vanish, as if the counters were not working. Then, when the satellite dropped back below 2,000 km, the counting resumed. The interpretation, confirmed by testing duplicate counters in the laboratory, was that the G–M counters on board were being over-loaded—which meant that the counting rate had to be at least 15,000 times larger than the usual CR rate. This was confirmed by Sputnik III and Explorer IV in 1958, both by then carrying counters designed to cope with the greater particle arrival rates.

The satellite data also showed that the intensity of this new radiation was correlated with magnetic but not geographic latitude; thus, like cosmic rays, the radiation had to be controlled by the geomagnetic field. Later space probes, some in orbits relatively close to Earth but others traveling out to distances of several Earth

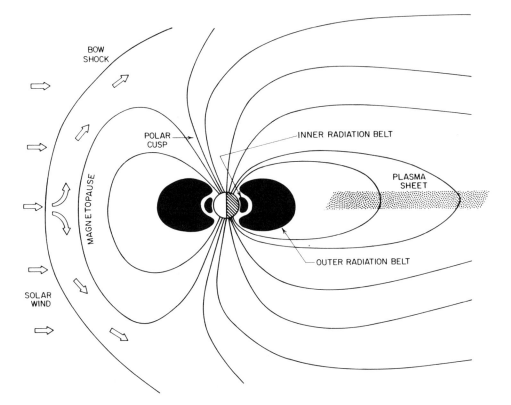

BOW
SHOCK

POLAR
CUSP

INNER RADIATION BELT

PLASMA
SHEET

MAGNETOPAUSE

OUTER RADIATION BELT

SOLAR
WIND

Figure 3.6   *Regions of trapped radiation around the Earth. These belts are controlled largely by the Earth's magnetic field, but the influence of the solar wind distorts the outer belt, compressing it on the sunward side and extending it on the side away from the sun. (Courtesy of J. A. Van Allen. Reprinted with permission from James Van Allen, "Radiation Belts," in* Encyclopedia of Physics, *ed. Lerner and Trigg, © 1981, VCH Publishers, Inc., New York, N.Y.)*

radii, provided the data to complete the picture. It soon was seen that the Earth was surrounded by two gigantic radiation belts (Figure 3.6) in which large numbers of high-speed electrons and protons were moving, trapped in orbits that, only twenty years earlier, had seemed to be just a mathematical curiosity.

Why didn't the first two Sputniks discover these radiation belts? Although Sputnik II, unlike Sputnik I, carried counters, the shape of its orbit and telemetry restrictions provide the answer. Data were telemetered back to Earth only when the satellite was over friendly territory, and at those times the satellite was not penetrating the dense parts of the trapped particle region. If the Russians had telemetered their data more continuously, in a form that could be understood by receiving stations elsewhere, we would probably now be referring to the Vernov radiation belts rather than the Van Allen belts. Van Allen announced his discovery in May 1958, and until then there had been absolutely no hint that the Russians had noted anything unusual in their counting rates.

Before examining this trapped radiation, it is useful to take a closer look at the geomagnetic field to see how it controls the movement of charged particles. The field around a magnet can be pictured as filled by a family of lines. If a small magnetic compass is used to survey the field of a large magnet, the lines of force will follow the direction of the test-compass at different places (as with a magnet held under a sheet of paper on which iron filings are sprinkled). The term "lines of force" was introduced by Niccolo Cabeo in 1629, but it was Michael Faraday, a scientist of great insight and inventiveness working in London's Royal Institution in the nineteenth century, who developed a quantitative model of magnetic fields—a visualization still of great use and found in most textbooks.

It is also helpful to review the quantities and units that describe the energies of CR particles. The laws governing moving objects and particles were set out in their modern form by Isaac Newton in 1687 in his great treatise the *Principia,* and his methods are still used today. The concept of speed is familiar from everyday experience, but calculating how a particle responds to forces (such as the Earth's magnetic field) requires that the particle's mass be taken into account. Two auxiliary quantities that combine speed and mass are *momentum, mv,* and *kinetic energy, $\frac{1}{2} mv^2$,* where m denotes the particle's mass and v its speed (or velocity).

These expressions for momentum and energy must be modified when the particle's speed is greater than about one-tenth of the speed of light (c), to take into account effects first recognized by Albert Einstein in 1905. Although v must always be less than c, as postulated by Einstein, the particle's mass and kinetic energy can increase without limit. Consequently, it is much more convenient to describe particles by their (kinetic) energies than by their speeds. The usual units for energy are electron volts (eV) and its multiples of one thousand (keV), one million (MeV), and one billion (GeV). (The European GeV is preferred to BeV for 10^9 eV, to avoid confusion. In the United States "billion" denotes one thousand million, whereas in European usage it is one million million.) This terminology dates from the early practice of using high voltages to accelerate cathode rays. The electron volt is an extremely small unit of energy, but quite appropriate to the scale of atomic particles. If we used electron volts for everyday objects, we would find that the kinetic energy of a baseball thrown at a speed of 30 miles per hour could be expressed as 8×10^{19} eV. Yet the highest-energy CR particles weighing 10^{26} times less, have about twice as much kinetic energy because of their very high speeds.

Table 3.1 lists a few representative values for the speeds and kinetic energies of protons (hydrogen nuclei) and electrons that are encountered in CR studies; it can be seen that the energy increases even though the particle's speed approaches its limiting value, *c*. An alpha particle from the radioactive decay of a radium or uranium nucleus has an energy of around 5 MeV, and CR energies range from tens of MeV to more than 10^{20} eV.

In the Van Allen radiation belts we find electrons with typical energies of a few MeV and protons of tens of MeV up to a few hundred MeV. A particle with an energy in this range will move around a line of force in a spiral path that becomes more tightly coiled as it comes closer to the Earth. Eventually the path becomes momentarily parallel to the Earth's surface, and then the particle starts to move back up. This bounce or reflection is governed by the field strength and the particle's momentum; the field acts like a magnetic mirror. After reflection the particle spirals up and loops over to a corresponding mirror point over the opposite hemisphere, where the process is repeated. A trapped particle will bounce back and forth repeatedly between hemispheres while at the same time drifting in longitude. A similar magnetic effect is employed in experiments designed for hydrogen fusion (thermonuclear) reactions: a carefully shaped magnetic "bottle" confines a hot ionized gas (a *plasma*) so it cannot touch (and vaporize) the walls of its containing vessel.

The bouncing can continue indefinitely for trapped particles in the Earth's radiation belts, until some new influence intrudes. Because the magnetic field is not perfectly regular, the drift motion is actually very complex, and some particles can transfer to different field lines that will guide them down into denser regions of the atmosphere where the particles will be lost through collisions. Alternatively, particles can be lost by collisions among themselves or by distortions of the magnetic field during times of increased solar activity. In all, the typical residence time of a proton in the inner radiation region is measured in years, and that of an electron varies from about a year in the inner belt to only a few days in the outer regions.

Table 3.1 *Particle speeds expressed as fractions of the speed of light*

Kinetic energy	Electrons	Protons
1 keV	0.063	0.0015
1 MeV	0.94	0.046
100 MeV	0.999987	0.43
1 GeV	0.99999987	0.88

The structure of trapped radiation belts is sensitive to solar influences by way of the solar wind; a typical depiction is shown in Figure 3.6. The inner belt, most heavily populated by protons, is about 1.5 Earth radii out, and the outer belt, containing most of the electrons, is about 4.5 Earth radii. There are, of course, no sharp edges to these regions; as the figure shows, the particles are spread around the Earth in a doughnut-shaped region in which the filling varies in density. In the inner belt, the number of protons peaks at about 15 per cubic meter. A few trapped helium nuclei have also been detected, around one for every 5,000 protons, and a smaller number of oxygen ions.

At its peak, the stream of the lower-energy inner-belt protons (those with energies above about 0.5 MeV) has about 200 million particles crossing an area of one square centimeter each second, while the corresponding figure for higher-energy protons (above about 15 MeV) is about 2,000 times smaller. For electrons, the inner-belt intensity peaks at about $60/cm^2$ sec, and then drops with distance only to increase again to another maximum of 1 million/cm^2 sec in the outer belt.

The numbers given here for the populations of the trapped particles are useful as guides, but those populations do vary greatly. The structure of the outer regions of the Earth's field is sensitive to the magnetic field carried by the solar wind, and that field in turn is subject to the variable level of solar activity, which follows a general 11-year cycle with sporadic additional outbursts. In the region where the solar wind sweeps by the Earth, the collision of the two magnetic fields creates a complex boundary layer. The Earth's magnetosphere is somewhat compressed on the sunward side and is pulled out considerably more on the opposite side of the Earth. The inner radiation belt is little affected by these changes, but the distant regions of the outer belt are much more responsive.

The basic correctness of this model of the Earth's field with its trapped particles has been confirmed by experiments with high-altitude nuclear explosions. These tests were carried out from 1958 to 1962, before growing public concern over radioactive fallout led to the adoption of the first test ban treaty. In the test explosions, large numbers of high-energy electrons were released and their progress followed as they spiraled along the magnetic field lines that passed through the locations of the test shots. In a particularly dramatic demonstration, the Argus shot detonated over the Johnston Islands in the Pacific produced an aurora later seen over Samoa, about 2,000 miles to the south. These two locations are at magnetic conjugate points (symmetrically placed north and south of the magnetic equator), and electrons followed the field lines until they reached a mirror point over Samoa where atmospheric collisions created the aurora, a rare phenomenon at tropical latitudes.

Another magnetosphere with a population of trapped particles was discovered in 1955, when Jupiter was identified as the source of intense and variable radio signals. These radio emissions, studied from the Earth and from the Pioneer and Voyager long-range probes, revealed that Jupiter possesses a magnetic field much stronger than the Earth's, with large numbers of trapped electrons and heavy ions. Jupiter is about five times farther from the sun than the Earth is, so solar-wind effects on Jupiter's magnetosphere are different in scale. The phenomenon of radio emission by fast electrons traveling along curved paths in a magnetic field (*synchrotron radiation*) was first elaborated for astrophysical settings largely by the outstanding Russian theorists Iosef Shklovsky and V. L. Ginzburg. A curiosity of Jupiter's emissions is that they are strongly influenced by Io, the innermost of the four moons discovered by Galileo in 1610. The cause of Io's influence has not yet been found, but it should be noted that Io is well inside Jupiter's trapped electron region, in contrast to our moon, whose orbit places it about ten times farther out than our outer radiation belt.

Our understanding of the sources of the Earth's trapped particles is far from complete. Soon after Van Allen's discovery, a plausible explanation attributed the trapped particles to primary CR collisions in the Earth's atmosphere. In many of these collisions, neutrons are ejected from atomic nuclei. With a half-life of about 11 minutes, each neutron decays to produce a proton, an electron, and a neutrino. Because neutrons carry no electric charge, they are unaffected by the geomagnetic field and travel directly away from the sites of their release until they decay. In this way they can reach any point in the magnetosphere and produce electrons and protons in places they could not otherwise have reached. Some of these electrons and protons will have the right speeds and directions of travel to allow them to be trapped. This mechanism, however, cannot account for all of the trapped particles apart from the most energetic protons (with energies above about 50 MeV) in the inner belt. Recent theory places the origin of most particles in a combination of solar wind, CR interactions, and the ionosphere. Through processes that we understand only partially, slow particles are accelerated by changing electric and magnetic fields. A similar situation exists for the explanation of the auroral particles.

Geophysical studies have thus provided (in the latitude and East-West effects) the basis for identifying the primary CR as electrically charged particles rather than the (electromagnetic) photons that Millikan had initially preferred. When the IGY monitoring of cosmic rays led to the entirely unanticipated discovery of trapped particles, it was again geomagnetic theory that provided the framework for a model.

Particles from the Sun

The Earth's location in the solar system preserves a delicate thermal balance. If we were much closer to the sun our atmosphere would lose its essential lighter gases, and if we were much farther away life either would have evolved differently to cope with freezing temperatures or perhaps would never have developed at all. The sun's energy, originating from fusion reactions deep within, diffuses to the surface and radiates in all directions, with less than one part in a billion striking the Earth. Because the solar interior is so dense, the energy takes more than a million years to travel from the core to the surface, but then only another 8 minutes to go the additional 150 million km to the Earth. (This distance, equivalent to 93 million miles, is often used as a measure of distances within the solar system and is called the astronomical unit or AU.)

The sun maintains its radiant output at a generally steady level, a baseline known as the *quiet sun*. At times an *active sun*, with an increased number of sunspots, can create on Earth a disruption in long-range radio transmission. Other indications of increased solar activity will be discussed later in this chapter.

More than electromagnetic energy is departing from the sun. The bright *photosphere* that we see has a temperature of around 6,000 °K, but the overlying *corona* (spectacularly visible during solar eclipses) has a temperature range that extends to over a million degrees. At this temperature some ionized gas constituents are moving at speeds great enough to escape from the solar gravitational attraction, flowing out through the solar sytem as a solar wind. The flow amounts to about 300,000 tons each second, but this seemingly enormous quantity still amounts to only one part in 10^{14} of the solar mass each year.

Although it might seem incredible that the sun's loss of hundreds of thousands of tons each second could escape our notice, that was indeed the case until as recently as the 1960s, when the existence of the solar wind was firmly establishd. The solar wind is now easy to detect with particle detectors carried on satellites well above our atmosphere, but before the space age the only indications of its presence were less direct. There had long been speculation that the sun influenced our weather, and of course astrology was based on the supposition that events on Earth are controlled by the sun and other heavenly bodies merely by their positions. The first scientific intimation of a solar influence came in 1859, when R. C. Carrington, an English amateur astronomer, observed a *solar flare*, an intense local brightening on the face of the sun close to a sunspot. At the same time, monitors of the Earth's magnetic field recorded changes from their usual values. Such events are termed *magnetic disturbances* or *storms*. The nature of any physical way in which the sun could control these phenomena on Earth was quite unknown. Carrington also noted that about 12 hours after the flare a major magnetic storm accompanied by auroras was seen in both hemispheres. The possible connection between solar activity and auroras was a subject of continued investigation, but the scattered speculations were inconclusive.

A related solar phenomenon also stimulated the search for possible solar-terrestrial connections. On many occasions spots had been observed on the disk of the sun, but they were invariably ascribed to the passage of planetary bodies such as Mercury or Venus in front of the sun, since people were reluctant to challenge the presumed unblemished nature of the sun. After Galileo's invention of the astronomical telescope in 1609, however, this interpretation had to change. One of his early discoveries was that the spots were truly attached to the sun. Continued observations showed that the spots moved across the sun's face as though the sun itself were rotating; confirmation came when spots were seen to disappear at one limb of the sun and reappear round the opposite limb about two weeks later, implying a 27-day rotation period. (The true rotation period is closer to 25 days; the longer apparent period is a result of the fact that the observations are being made from the Earth, which is not at rest but moving in its orbit around the sun.)

The number of sunspots visible at any time is highly variable. In 1877 R. Wolf of Zurich compiled records of sightings dating back to 1610 and found that sunspots occurred in cycles, their numbers increasing and decreasing over intervals of about 11 years (Figure 4.1). The Zurich sunspot number, still our most direct indicator of solar activity, has a periodicity similar to that of geomagnetic disturbances, and it was natural to think that the latter might be related to the sunspot cycle. No physical link could be found, however, and the connection remained speculative. No less a figure than Lord Kelvin argued, in 1892, that "we may be forced to conclude that the supposed connection between magnetic

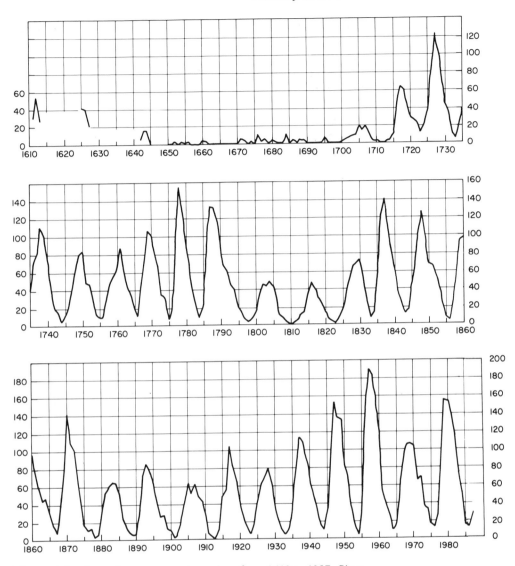

Figure 4.1 *Annual mean sunspot numbers from 1610 to 1987. Since about 1710 there has been a general 11-year periodicity, but with different numbers of sunspots seen in successive maxima. For the years preceding 1700, the 11-year periodicity has not been as firmly established. Sunspot observations were not routine before the 1609 invention of the telescope and were still sparse for many years subsequently. However, the paucity of sunspots during the years 1650 to 1700 is considered well established, and this period is now known as the Maunder Minimum. It has been correlated with many climatic effects, including low temperatures worldwide, early frosts, increased glacier sizes, and late blooming of cherry trees in the royal gardens in Kyoto. The underlying cause of this solar behavior is not known. (Courtesy of John A. Eddy, High Altitude Observatory.)*

storms and sunspots is unreal, and that the seeming agreement between the periods has been a mere coincidence." A few years later C. A. Young, a Princeton astronomer, expressed a more cautious judgment: "From the data now in our possession, men of great ability and laborious industry draw opposite conclusions" (*The Sun,* pp. 169 and 162).

Additional and more persuasive evidence for the extended influence of the sun accumulated from three rather different lines of research. First, the 27-day periodicity in geomagnetic activity persisted over many years. Over a short period one might be willing to consider the similarity of 27-day cycles as accidental, but not when the effect persists for 30 years and covers nearly 400 cycles, with the average length so similar to the solar rotation period.

Additional support came from CR studies, but in an unanticipated way. Soon after the discovery of cosmic rays, a search began for variations in their intensity, not only with altitude but also with time. The early records of CR intensity showed large variations, up to 15 percent, but there was no consistency between different measurements. More careful work, together with improved corrections based on better knowledge of CR processes in the atmosphere, proved that daily variation was itself variable, but on the average less than 0.5 percent. The modern era of CR monitoring started in 1936 when Scott Forbush, of the Carnegie Institution in Washington, set up a network of stations with recording ionization chambers in Greenland, the United States, Peru, and New Zealand. Forbush found that the intensity of cosmic rays reached a maximum each day, usually around 3:00 P.M., and then declined to a minimum around 12 hours later, with the swing from maximum to minimum being only about 0.15 percent on average. By 1957 Forbush had sufficient data to conclude that there was a correlation between long-term CR variations and solar activity, but an *inverse* one: over the 11-year solar cycle, when sunspot numbers increased, CR intensity *decreased* (Figure 4.2). Finally, also in 1957, H. V. Neher of the University of Minnesota, who had been conducting high-altitude balloon flights each year near the magnetic pole at Thule, Greenland, discovered that low-energy CR protons were significantly more numerous when solar activity was minimal.

These varied observations led to the surprising conclusion that solar activity, as indicated by the 11-year sunspot cycle, was not producing more cosmic rays but rather was somehow preventing galactic cosmic rays from reaching the Earth. Quite unexpectedly, the last piece in the puzzle came from studies of comet tails.

Comets occupy a distinctive place in astronomical folklore. Their unpredicted appearances used to be taken as omens, and Halley's comet is depicted in various early works of art such as the

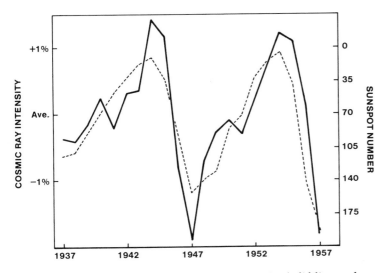

Figure 4.2 *Correlation between cosmic ray intensity (solid line, scale at left) and sunspot number (dashed line, scale at right) through two solar cycles. Increased solar activity produces a decrease in the cosmic ray intensity at Earth. (Adapted from Scott Forbush,* Journal of Geophysical Research *63 [1958]: 657, © American Geophysical Union.)*

1066 Bayeux tapestry and Giotto's 1304 fresco in the Scrovegni Chapel in Padua. The return of this comet in 1986 showed that these objects can still arouse great interest. The often spectacular comet tails always point away from the sun, so they will trail behind a comet as it approaches the sun and flow ahead of the comet as it sweeps away from the sun. How the tail appears to us depends on the relative positions of sun, Earth, and comet; occasionally a comet may even seem to have two tails pointing in opposite directions.

For some time it has been known that comets generally have two tails (Figure 4.3). Type I tails are clumpy and straight, whereas Type II tails are curved and more uniform in appearance. The tails are made of material that evaporates from the comet as it warms during its approach to the sun. Spectroscopic analysis of the light from the tails has identified their constituents. Type I tails are composed of gaseous ions such as CO^+ (carbon monoxide) and OH^+ (from the breakup of water vapor molecules); Type II tails, in contrast, contain many dust particles that were originally frozen into the icy comet but were then released as the binding ice evaporated. Under the pressure of solar radiation, the dust particles get blown away from the comet and can be seen by the sunlight they reflect. The ions are much smaller than the dust particles and hence far less affected by solar radiation.

Figure 4.3
*Comet West, photo-
graphed in March
1976. The separate
tails can be clearly
seen. (Photograph
courtesy of Lick Ob-
servatory.)*

Between 1951 and 1957, Ludwig Biermann of Gottingen studied Type I ion tails. In photographs taken over many days, Biermann followed the movement of dense clumps of ionized gas away from the head of the comet and deduced the changes in speed of these clumps. From these observations he inferred the magnitude of the force that had to be responsible and suggested that these effects could happen if the sun were continually emitting particles whose collisions with ions in the tails provided the driving force. Biermann was able to deduce the speed and the numbers of particles in this solar stream.

In 1958, a seminal paper was published by Eugene Parker of the University of Chicago. Drawing on Biermann's work, Parker set out a detailed mathematical model of the flow of solar particle radiation, associating it with other features of the sun such as its magnetic field, corona, and rotation. Parker's introduction of the concept of a solar wind was "received with a great deal of skepticism, even disbelief" (as the NASA astronomer Jack Brandt described it), but the rest, as they say, is history. Parker's magnetohydrodynamic model has been refined, and many features have been verified by satellites and long-range probes.

The first direct observation of solar wind particles came from the MIT detectors on Explorer 10 in 1961, with many additional observations made during the following years. Near the orbit of the Earth (at a distance of 1 AU from the sun), the solar wind moves with an average speed of 450 km/sec (about 300 miles/sec) and contains about 5 protons per cubic centimeter—which means a flow of more than one hundred million particles per square centimeter per second. The quantities vary considerably, and streams of particles sometimes move faster than 1,000 km/sec.

The solar wind is composed of particles with both positive and negative charge (protons and electrons), but because they occur in equal numbers, the stream is electrically neutral. The high temperatures (over one million degrees at the sun and half that by the time the solar wind reaches the Earth) ensure that the gas remains highly ionized. Within the stream is a magnetic field, so there is a magnetically complex region where the solar wind flows by the Earth and encounters the Earth's own magnetic field. The latter is distorted: compressed on the side toward the sun and drawn out on the opposite side. As a result, the Van Allen regions of trapped particles are also distorted (see Figure 3.6), as discovered and mapped by spacecraft. Because of the sun's rotation, the solar wind does not simply stream out along straight radial directions like the spokes in a wagon wheel but follows curved spiraling paths, in what is often called the garden-hose effect. The Earth thus encounters the solar wind coming from a direction slightly ahead of the direct Earth-sun line. The solar wind has been detected as far out as 40 AU from the sun by the Pioneer 10 spacecraft. At much greater distances, 50–100 AU, it is thought that the solar wind terminates abruptly in a "shock," a complex boundary between the interplanetary and interstellar regions.

What do we know of the processes that produce the solar wind? Ultimately, the energy comes from the sun's nuclear reactions, which sustain the high temperatures in the corona. Recent satellite observations of the sun, in the X-ray and ultraviolet regions, have shown *coronal holes,* regions of low density where the magnetic field of the sun is carried out with the escaping streams of particles. High-speed streams appear to be associated with coronal holes near the sun's polar regions, but slower streams come from all regions. The whole picture is strongly influenced by the sun's rotation and its intricate magnetic field. Thus the general qualitative features of solar wind production seem to be understood, but the description is mathematically complex and much detail remains to be clarified.

The solar wind is a persistent though variable part of the quiet sun's output. Solar cosmic rays, however, represent one of several short-lived manifestations of the active sun (Figure 4.4). The solar wind and solar CR are composed of the same types of particles, electrically charged constituents of the outer solar layers. If you examined an individual particle, you could not tell whether it belonged in the solar wind or among the solar CR. Both groups are emitted sporadically, and it is difficult to correlate the variations with the solar cycle. What does distinguish these two groups of particles are their numbers and speeds: the cosmic rays have speeds and energies vastly greater than their solar-wind relations. A typical solar-wind proton will have a speed of 500 km/sec (about 0.2 percent of the speed of light) and an energy of 3,000 eV, while solar CR have energies measured in millions of electron volts and speeds above 15,000 km/sec (about 1/20 of the speed of light).

In the late 1950s, around the time of the IGY, several independent observations pointed to the clumpy emission of high-energy solar CR. Earlier Forbush had noted an occasional decrease in the rates of his ground-level ionization chambers, but in February 1956 his station at Huancayo in Peru showed a huge increase in the CR intensity (as in Figure 4.4). At this location, near the magnetic

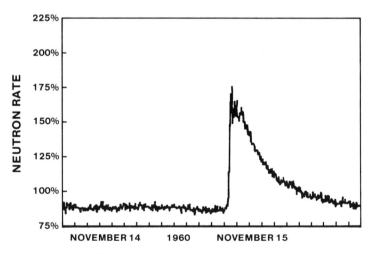

Figure 4.4   *Sudden increase in cosmic ray activity associated with a large solar flare in November 1960, as shown by a sea-level neutron counter at Deep River, Canada. (Adapted from J. F. Steljes, H. Carmichael, and K. G. McCracken,* Journal of Geophysical Research 66 *[1961]: 1363, © American Geophysical Union.)*

equator, particle energies above 13 GeV were needed to get close to the Earth. At the same time, D. K. Bailey (of Page Communications Engineers and the National Bureau of Standards) noted major changes in the attenuation of radio signals. Absorption of the solar ultraviolet radiation maintains the degree of ionization (or the numbers of electrons) in the ionosphere, 50 km or more from the Earth's surface. These ionized layers, in turn, control long-distance radio transmission. Changes in radio transmission thus provide a good way to monitor changes in the ionosphere, and Bailey's measurements pointed to the arrival of large numbers of ionizing particles.

Earlier high-altitude balloon flights had uncovered flare-related particles whose energies were too low to produce secondary effects detectable at sea level. The first clear observation of solar-flare protons was obtained by Kinsey Anderson, in Van Allen's department at the State University of Iowa, during a 1957 balloon flight at Fort Churchill near Hudson Bay. Other high-altitude CR observations together with the monitoring of radio transmission showed that radio blackouts occurred over large areas when many low-energy solar protons were channeled into the polar regions by the Earth's magnetic field. In the jargon of the field, the solar proton events produced *polar cap absorption* (PCA).

In a PCA event, the arrival of solar ultraviolet and X-ray photons increases the number of ionospheric electrons and inter-feres with radio communication, which, though a nuisance, does not damage radio receivers; transmission returns to normal as additional electrons reattach to atoms. We now have the ability to simulate something like a PCA, but the consequences are poten-tially serious. The detonation of a large hydrogen bomb above the atmosphere creates an intense fireball of gamma rays. From a typical altitude of 100 km, many of these photons enter the atmo-sphere and ionize numerous atoms. The sudden release of electrons triggers an intense electromagnetic pulse (EMP) that propagates through the atmosphere and can damage transistorized circuits much as an overload in a house can blow a fuse. The deliberate use of a nuclear bomb to incapacitate the communications network of an enemy is a cause of concern to the military. Shielding against EMP is not impossible but does carry a weight penalty that can be a serious problem for airplanes and satellites.

Large solar proton events directly related to major solar flares can be monitored optically. In a major flare, this light intensifies for about a minute and then rapidly dims. Similar flaring is generally detected at short wavelengths such as ultraviolet and X-rays. Other visible signs of solar activity can have an awesome scale. Figure 4.5 shows the giant *solar prominence* of June 1946. During an interval of only 98 minutes, the heated gas rose (at speeds of around 100

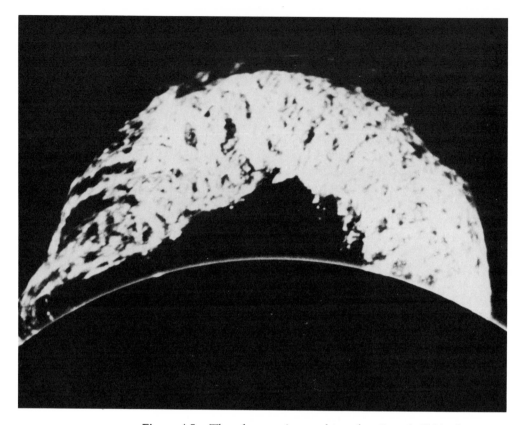

Figure 4.5   *The solar prominence observed on June 4, 1946, photographed through a red filter to show hydrogen emission. The gas temperature varied between 25,000 and 100,000°K. Within an hour this prominence grew to become nearly as large as the sun itself; its shape was strongly controlled by the local magnetic field. (Photograph courtesy of High Altitude Observatory.)*

km/sec) to a distance comparable to the solar radius, or about 100 times the radius of the Earth. The role played by the sun's magnetic field in confining the motion of the charged particles in the glowing gas can be seen clearly in the sharp boundaries of the prominence and its looping shape.

The travel time from the sun to the Earth for electromagnetic radiation is 8 minutes, so the ionizing ultraviolet and X-radiations arrive as the flare is being seen and promptly increase the number of electrons in the ionosphere. Radio transmission is then noticeably affected. The charged particles travel fast but still much slower than the light; their arrival at Earth follows after a delay that depends on their own speeds and on the solar wind structure at the time. Typical delays are measured in hours, as Carrington and others had seen (but not understood). The geophysical effects on the iono-sphere and radio transmission are too complex to pursue here, but the main features are now well established.

The occurrences of major flares are unpredictable, but their frequency follows the general sunspot cycle. High-altitude rockets have been used to detect the flare particles at times of expected increased solar activity. The advantage of rockets is the speed with which they can reach full altitude, far faster than balloons or sat-ellites, which require protracted preparation and cannot be held ready for extended periods. On one occasion Carl Fichtel and Don Guss, of the NASA Goddard Space Flight Center, shot off rockets to observe the large solar flare of November 1960. Their rockets reached a peak altitude of 400,000 feet in a few minutes and were above 100,000 feet for about 5 minutes. Many solar protons were detected, as well as heavier particles, and thus these researchers were able to obtain a value for the hydrogen/helium ratio for solar CR. The presence of helium in the solar atmosphere had been known since its discovery by Lockyer in 1868, but estimating its abundance had been very difficult. The CR observations provided a reliable and new method.

Because CR particles are identified by their charges, the list of their abundances is usually called the *charge spectrum*. The charge spectrum of the solar cosmic rays, listed in Table 4.1, is well established. The abundances, at least for energies above 10 MeV, vary by less than a factor of three between different events, and it appears that we are observing a representative sample of coronal material that has been accelerated.

Another characteristic of solar CR is their energy spectrum. Cosmic rays do not all have the same energy, and so do not all travel at the same speed. The number of particles with energy E (in MeV) is proportional to E^{-a}. The minus sign indicates that there are fewer particles with larger values of E, and the value of a is usually in the range 2.5–3.

The intensity or flux of cosmic rays is the number of particles that traverse a standard area (such as one square centimeter) each

Table 4.1 *Average solar cosmic ray abundances (relative to hydrogen = 100)*

Nuclei	Abundance
Hydrogen	100
Helium	14
Lithium, beryllium, boron	<0.004
Carbon, nitrogen, oxygen	0.23
All heavier nuclei	0.037

second. (Strictly speaking, intensity and flux are different measures, but this fine point can be ignored here.) During large flare events the solar CR intensity can exceed 100,000 particles/cm^2 sec, with the total energy amounting to 10^{31} ergs. This is roughly equivalent to the energy released by 2.5×10^8 hydrogen bombs, but still only equal to the solar e–m radiation emitted in 1/300 of a second.

The acceleration of CR particles takes place in the magnetic field associated with sunspots, through magneto-hydrodynamic processes. The mechanisms involved are probably similar to those that produce galactic cosmic rays, to which I turn in the following chapter. Although most of the experimental methods for detecting galactic CR are similar to those for solar CR, the subject of galactic particles leads to a whole new range of astrophysical areas.

Cosmic Rays in the Galaxy

Cosmic ray studies extend far beyond the familiar region of the solar system to the galaxy, where vastly greater distances and different physical processes are encountered. The nature of the Milky Way was the subject of speculation and myth for centuries; then in 1609 Galileo's invention of the astronomical telescope changed astronomy forever, freeing it from the limits set by the human eye. Galileo turned his glass in all directions and described his findings in his book *The Starry Messenger*, recording with surprise and exhilaration the nature of the Milky Way (*Discoveries and Opinions of Galileo*, p. 49):

> I have observed the nature and material of the Milky Way.
> With the aid of the telescope this has been scrutinized so
> directly and with ocular certainty that all the disputes
> which have vexed philosophers through so many ages
> have been resolved . . . The galaxy is, in fact, nothing but
> a congeries of innumerable stars grouped together in clus-
> ters. Upon whatever part of it the telescope is directed, a
> vast crowd of stars is immediately presented to view.
> Many of them are rather large and quite bright, while the
> number of smaller ones is quite beyond calculation.

From the lack of any perceived relative motion among the stars, astronomers long ago made the correct inference that they were much farther away than the obviously wandering planets. Just how much more distant proved difficult to determine, and it was only as recently as 1836 that the first reliable distance measurements were made independently by Friedrich Wilhelm Bessel in Germany, Otto Struve in Russia, and Thomas Henderson at the Cape of Good Hope. Since then larger telescopes and more reliable measurements of stellar distances have led to the picture we have high-energy cyclotrons, but these nuclei are radioactive with short half-lives and do not occur in nature.

An electric force attracts electrons to their parent nuclei; this force also holds atoms together in molecules, the result of sharing some of the outer electrons of each atom. Whether a conglomeration of molecules or atoms is solid, liquid, or gas depends on the properties of those atoms or molecules and on the temperature. At all temperatures atoms and molecules are moving, even in solids. This internal motion becomes more vigorous as the temperature rises, until a solid melts to become a liquid and the liquid in turn evaporates to form a gas. Raising the temperature causes molecules in a gas to collide with increasing violence until they are torn apart into their constituent atoms. Further increases in temperature lead to more frequent and violent collisions until eventually electrons get knocked off the atoms and the gas becomes a *plasma,* a cloud of positively charged ions or nuclei and negatively charged electrons.

The temperatures at which these changes occur differ from one type of atom to another, and also depend on the pressure. For example, at normal atmospheric pressure, H_2O molecules condense to form water at 100 °C and solidify to form ice at 0 °C. Hydrogen, usually thought of as a gas, is liquid at temperatures below -254 °C. The surface of the sun has a temperature of about 6,000 °C, and in that region about one hydrogen atom in a hundred thousand is ionized. Complete ionization of hydrogen requires a temperature of around 10,000 °C, which is easily the case in the outer regions of the solar atmosphere (the corona) where the temperature exceeds one million degrees.

Because of this diversity of physical states, different methods of analysis must be used in compiling lists of abundances of different elements, depending on whether they are located on Earth, on the sun, or on other stars. Allowance must also somehow be made for the bulk of the universe beyond our immediate reach. Chemical analysis can be used for backyard dirt, the atmosphere, and the oceans, but what about the Earth's interior? Even the deepest mines go down only about two miles below the surface, a tiny fraction of the Earth's radius of 4,000 miles. By examining the sun's light in a spectroscope (Figure 5.2), we can identify different types of atoms from their characteristic spectral features, but what about the bulk of the sun that is far below the glowing photosphere?

There is no simple way to provide a definitive tabulation of elemental abundances. An obvious start is with materials that can be brought into laboratories, followed by those that can be examined only from a distance by using astronomical techniques. To proceed farther one must resort to models, theoretical pictures that combine observed features with plausible assumptions. From such models one can deduce the overall composition of the Earth,

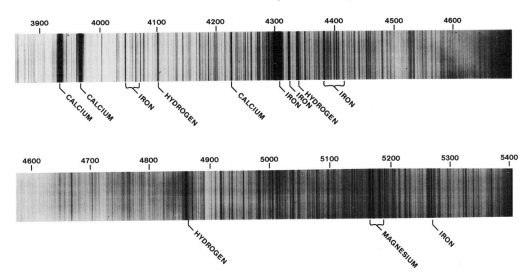

Figure 5.2 *A section of the solar spectrum covering the blue through green wavelengths (top left to lower right). Some of the prominent lines produced by different types of atoms have been labeled. (Courtesy of Mount Wilson and Las Campanas Observatories, Carnegie Institution of Washington.)*

the moon, the sun (Table 5.1), and more distant bodies; further observations may necessitate adjustment to the models.

Apart from materials found on Earth, three significant sources of material are available for analysis in laboratories: meteorites, dust particles, and samples returned from lunar missions. Countless objects, ranging in size from tons down to micrograms or smaller, orbit the sun, and their collisions with the Earth are matters of chance. Those that enter the Earth's atmosphere do so at

Table 5.1 *Solar composition*

Element	Atomic number (Z)	Number of atoms (relative to hydrogen)	Fraction of total mass
Hydrogen, H	1	1.00	0.784
Helium, He	2	6×10^{-2}	0.198
Carbon, C	6	3×10^{-4}	3×10^{-3}
Nitrogen, N	7	2×10^{-4}	2×10^{-3}
Oxygen, O	8	6×10^{-4}	8×10^{-3}
Neon, Ne	10	1.3×10^{-4}	2×10^{-3}
Iron, Fe	26	8×10^{-6}	4×10^{-4}

Source: Adapted from E. G. Gibson, *The Quiet Sun* (Washington, D.C.: N.A.S.A., 1973), p. 72.

high speed and heat up. The smallest objects burn up completely, but the largest (the meteorites) can survive to reach the ground, though they lose much of their outer material as they plunge through the atmosphere. The composition of these objects, large and small, gives clues to the nature of the giant cloud of gas and dust from which the solar system is thought to have condensed about 4.5 billion years ago. Meteorites are classified as stony, iron, and stony-iron; one subclass, the carbonaceous chondrites, are regarded as the most primitive. Some micrometeorites, truly cosmic dust, have been detected from satellites and high-altitude research planes. Study of these tiny particles, typically no more than 1/40 mm in diameter, has given new insights into interplanetary composition and processes in the early solar system.

The third type of extraterrestrial material is lunar samples. During the six Apollo missions a total of 381 kg (838 pounds) of lunar material was returned to Earth, and the Russian Luna 16 and 20 unmanned vehicles brought back 130 g. Less than 10 percent of the Apollo material has thus far been distributed for analysis; the remainder is preserved at the NASA Johnson Space Center in Houston for future research.

In compiling a tabulation of elemental abundances, we must also include the material that is widely dispersed between the stars. That region is not totally empty but contains the interstellar medium (ISM), a thin gas composed of stray atoms and molecules with some dust particles. There are also some giant gas clouds in which the density is about a million times higher, with a rich variety of complex organic molecules. The interstellar region is a remarkably good vacuum: a typical cubic centimeter contains only a single hydrogen atom. In contrast to these low gas densities, the best vacuum achieved under laboratory conditions is about 10^{14} atoms per cubic centimeter. There are two ways of learning about the constituents of the ISM: some regions are dense and warm enough to emit measurable amounts of radiation, while others reveal themselves by their absorption of light from more distant stars whose spectra display characteristic features produced by the ISM.

From all these sources, the overall abundance of each element can be assessed. But an additional classification can be made that can lead to important insights into the operation of nuclear processes in the astrophysical setting. This classification is by *isotopes* within each element. Atoms were historically classified by their ability to combine (chemically) with other atoms. We now know that this property depends on the electrons surrounding each nucleus. Within each nucleus are positively charged protons and electrically neutral neutrons, held together not by any electrical force but by a much stronger nuclear force. Each chemical element is characterized by the number of protons in the nucleus, but many elements have nuclei with the same number of protons but different numbers of neutrons—the isotopes. The nucleus of the usual hydrogen atom is the proton, but a nucleus consisting of one proton and one neutron will still have only one orbiting electron. Chemically like hydrogen, this isotope is sometimes called heavy hydrogen but more often deuterium, and its nuclei are known as deuterons. The usual carbon atoms that we encounter are mostly carbon 12, ^{12}C, with six protons and six neutrons in each nucleus, but two other types also occur, ^{13}C and a radioactive isotope, ^{14}C, whose interesting properties are examined in Chapter 11.

It should be evident by now that compiling a table of "universal" or "solar system" elemental abundances against which to compare cosmic rays involves a complex array of resources: the data from solar and stellar spectroscopy, meteoritic and geologic data, and a considerable knowledge of nuclear theory (some theory is needed because not all of the isotopic abundances have been measured). The modern theory of stellar evolution includes calculations of nuclear reactions that steadily convert hydrogen to heavier nuclei, producing all the atoms in the periodic table. The general features of the nucleosynthetic processes are well understood and provide the basis for filling gaps left in experimental tabulations. The results, summarized in Table 5.2, are based on compilations by A. G. W. Cameron of the Harvard-Smithsonian Center for Astrophysics, who, probably more than anyone else, has continued to refine these tabulations over the past thirty years. The complete list includes estimates of abundances of all isotopes in addition to the element-by-element totals.

Table 5.2 *Solar system abundances*

Element	Relative number of atoms
Hydrogen	1.00
Helium	6.8×10^{-2}
Lithium, beryllium, boron	2.6×10^{-9}
Carbon, nitrogen, oxygen	1.2×10^{-3}
Neon	9.8×10^{-5}
Iron	3.4×10^{-5}
All heavier atoms	1.9×10^{-6}

Source: Adapted from A. G. W. Cameron, in C. A. Barnes, D. D. Clayton, and D. C. Schramm, *Essays in Nuclear Astrophysics* (Cambridge: Cambridge University Press, 1982), table 1, pp. 28–29.

Improvements in experimental methods have enabled us to measure the masses and thus identify the isotopes of at least the lightest CR nuclei. Knowledge of individual isotopic abundances among cosmic rays helps to identify the nuclear processes in which the particles were produced, for example, in stellar interiors or out in the interstellar space. The relative abundances of individual elements and some isotopes have by now been well established for cosmic rays, up to iron ($Z = 26$); beyond that the general trends, with some detail, have also been determined. In the following chapters I will make use of this knowledge as I try to reconstruct the history of the cosmic rays and identify their sources.

In describing galactic cosmic rays I will follow the pattern set with solar particles, beginning with their charge and energy spectra. With solar flare events, if the numbers of particles and their energies are known, one can calcualte the total energy and see, for example, how small a flare is in comparison with the sun's regular radiant output. A survey of the corresponding properties of galactic cosmic rays will make it possible to construct a model for their production and their travel (propagation) in the galaxy. But new physical processes—quite negligible in the relatively minuscule solar system—become important because of the galaxy's scale.

A perpetual problem in observing cosmic rays is their paucity, and the situation gets worse when one is examining heavier particles or those that are the most energetic. Confidence in intepretations is based on the reliability of observations and thus on the number of particles detected, the certainty of their identification, and the accuracy of energy measurements. Detecting more particles is quite simply a matter of using larger apparatus for longer times, and thus the equipment used in CR studies has evolved from electroscopes a few inches in diameter to satellite-borne hardware weighing many tons.

Today satellites are the obvious route: get above the atmosphere and stay there for as long as necessary. Electronic detectors can relay information back to Earth, where analysis can take place in the comfort of one's office. In some cases, as in the manned Skylab and the space shuttle, photographic plates and some newer types of track detectors (made of plastics) have been returned from orbit for processing and examination. But before satellites flew, and even long afterward, balloons have hoisted our detectors high enough to map the major features of cosmic radiation. The story of scientfic ballooning deserves a book to itself: the development of special plastics to contain the lifting gas; the role of independent companies such as General Mills, Raven, Winzen, and Schjeldahl, who designed, made, and flew the balloons, and the far-sighted contribution of the Office of Naval Research, which funded the balloon programs for many years. Ballooning continues, now coordinated by the National Scientific Balloon Facility in Palestine, Texas. Balloon loads have grown to tons, and volumes have increased to 30 million cubic feet. The science of balloon flying has become more sophisticated; early flights lasted only a few hours, but now these can be extended to many days by taking advantage of the seasonal reversals in the stratospheric winds. Anyone who has flown balloons will have a collection of stories about their flights (with good and bad endings). On many of our early flights we had a camera that took regular photographs of an altimeter and an Accutron watch to provide a record of the flight altitude. That watch survived a fall from 80,000 feet when the parachute failed— and was still running when the payload was found.

Hoisting a scientific load, whether by balloon or satellite, requires more than simply making sure the hardware functions as planned in a hostile environment. The particles detected and the ease of their identification depend on the location of the balloon flight or the orbit of the satellite. To understand the importance of these factors, it is necessary to probe a bit deeper into the method of identifying particles.

In Chapter 2 I described several methods of identifying particles, all based on measurements of the energy lost by a particle traversing the detector. This energy loss depends on the particle's charge and speed proportionally as Z^2/v^2. As the velocity (v) increases, the energy loss rate drops. There is a minimum loss rate, however, since the speed of a particle cannot exceed the speed of light in a vacuum. (This is a basic postulate of Einstein's special theory of relativity, well confirmed by experiment.) As energy increases, the energy loss rate goes through a minimum value and then slowly increases again.

The precise behavior of the energy loss rate at higher energies depends on the nature of the detecting material. A typical set of response curves is shown in Figure 5.3. To compare particles with different masses but the same speeds, we usually refer to kinetic energy per nucleon, MeV/n, where *nucleon* refers to the number of particles in the nucleus, regardless of whether they are protons or neutrons. A single proton and an alpha particle (containing two protons and two neutrons) that have the same speeds will have the same energy per nucleon, although the actual (total) kinetic energy of the alpha particle will be four times that of the proton.

Identifying a particle generally requires two independent measurements that depend in different ways on the particle's charge (Z) and speed. For particles with energies above about 1.5 GeV, the ionization loss rate (as seen, for example, in a scintillation

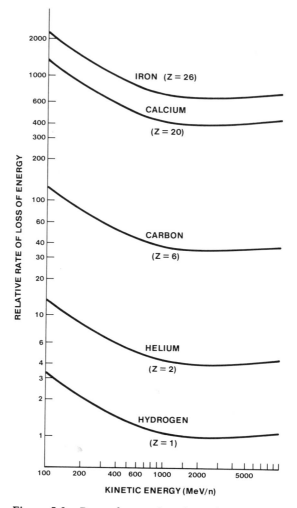

Figure 5.3 *Rates of energy loss shown for several particles with different charge (Z). The curves show the Z^2 dependence in their vertical spacing. From 100 to around 800 MeV/n, the energy loss rate decreases with the square of the particle's speed; after a broad minimum, the loss rate increases only slowly.*

counter) depends almost entirely on Z^2 (Figure 5.4). Provided the energy is in this range, a single measurement of ionization serves to identify the particle by its charge. This is where the Earth's magnetic field comes into the picture. When a detector is placed on a balloon over Texas, only particles with energies greater than about 1.5 GeV will be let in by the magnetic field, so one can concentrate on measuring ionization. For a balloon flight originating in Guam, closer to the magnetic equator, the magnetic cutoff energy is much

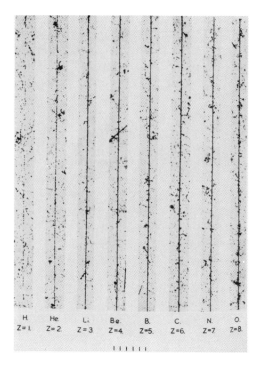

Figure 5.4
*Tracks of fast pri-
mary cosmic rays,
showing the depen-
dence of their ioniza-
tion on their charge.
The number of grains
and slow secondary
electrons (δ-rays) in-
creases with Z^2.
(Photograph courtesy
of Peter Fowler.)*

higher. On a flight from the northern United States or Fort Churchill in Canada, on the other hand, the cutoff energies are low. For
particles with energies below about 1.5 GeV, identification can be
made from several successive measurements of energy loss, sometimes in conjunction with deflection in a calibrated magnetic field
or with a speed-sensitive Čerenkov detector. During a satellite
flight the cutoff energy varies with each orbit, depending on the
spacecraft's position. The arrival time must be noted for each
particle so that corrections can be made later to allow for different
times spent by the satellite in regions of differing cutoff.

In Chapter 3 I described how in the 1930s scientists came to
understand the role of the Earth's magnetic field when the latitude
and East-West effects were discovered. These two observations
established the essential nature of most cosmic rays as positively
charged particles and ruled out Millikan's tenaciously held view
that the primary CR were gamma rays. Both direct and indirect
evidence pointed to protons and helium nuclei as the primary
particles.

The discovery of heavier particles came in 1948, as university
research resumed after the end of the war. Photographic emulsions
and cloud chambers were carried by balloons to altitudes above
90,000 feet, where they recorded particle tracks that were clearly
much more heavily charged than protons. In papers published

between 1948 and 1950, groups at the Universities of Rochester and Minnesota reported their discoveries. At the northern latitudes of the balloon flights, low-energy CR were seen, some so slow they came to rest in the emulsions. The Ilford and Eastman Kodak emulsions were not yet sensitive enough to record fast alpha particles, but many heavier particles could be identified, showing charges of $Z = 6$ or greater. The presence of alpha particles was established initially from tracks in a cloud chamber.

With these discoveries, the emphasis in CR research shifted to the charge spectrum—the determination of the abundance of each type of particle (Figure 5.5). The comparison of this spectrum with other astrophysical abundance tabulations remains a central focus of CR research. A brief review of some aspects of particle behavior will be useful as background to the interpretation of observations.

When two protons collide, they cannot be fragmented. Mesons can be created from the kinetic energy of the collision, and charges may be exchanged so that neutrons emerge instead of protons (with mesons taking care of the electric charge balance). The situation is more complex when heavy nuclei collide, for either the projectile or the target or both may be fragmented. This process can be described by three quantities. The *mean free path* (mfp) is the average distance a particle travels before it has a collision. Clearly, the mfp is large in interstellar space where target atoms are far apart, but it is much shorter in the atmosphere and becomes progressively shorter at lower altitudes as the air density increases.

The second quantity is the *cross section,* the effective target size (area), which depends on how many nucleons are in the nucleus. The volume of a nucleus is roughly proportional to the number (A) of nucleons (protons plus neutrons) it contains, and therefore the radius must be roughly proportional to the one-third power of A. The cross-sectional area will be proportional to the square of the radius and thus to the two-thirds power of A. The mfp and cross section are closely related and determine the frequency with which particles collide. A third quantity is needed to describe the types and numbers of fragments produced in each kind of collision.

Collisions take place both in interstellar space and in the atmosphere. In the early balloon flights, even those as high as 90,000 feet, the overlying atmosphere contained only about 1 percent of the total atmospheric atoms. In such a small amount of overlying atmosphere only about 10 percent of the CR protons will have collisions, but this is not the case for the heavier particles such as nuclei of carbon (C), nitrogen (N), oxygen (O), and iron (Fe), with their large cross sections. Many of these heavier particles will have collisions, and their fragments will include protons, alpha particles, and the nuclei of the light atoms, lithium (Li), beryllium (Be), and boron (B). When Li, Be, and B (the so-called L nuclei) were discovered in the 1948 emulsions, the immediate question was, were these true primaries that had survived through interstellar space and then through the atmosphere down to balloon altitude, or were they secondaries, fragments from the collisions of C, N, and O (the medium or M nuclei) or heavier nuclei? The importance of this question lies in the general scarcity of the L nuclei in nature, both on Earth and more broadly in the galaxy.

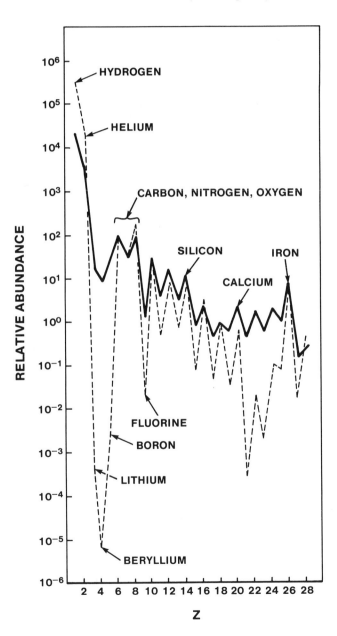

Figure 5.5
Charge spectrum of the cosmic rays, from Z = 1 *through* Z = 28, *as observed at the Earth (solid line), compared to the general solar system abundances (dashed line).*

The abundance of the L nuclei relative to the next heavier group, the medium or M nuclei (C, N, and O), became the focus of dispute in the 1950s, as Bernard Peters (at Rochester) maintained that the observed L nuclei were all atmospheric secondaries while Peter Fowler and his colleagues (in Bristol) claimed a significant presence of primary L nuclei, even after allowing for production in the atmosphere. The resolution of this controversy came from a 1958 balloon flight carrying emulsions to a new record altitude of 120,000 feet. The residual overlying atmosphere during this flight was less than one-quarter of 1 percent of the atmosphere, and the corrections for fragmentation were small enough that the results were beyond dispute. The flight location had been chosen in such a way that only relativistic particles would be detected, to reduce the uncertainties in particle identification. The group at the Naval Research Laboratory under Maurice Shapiro (who had been a student of Compton) reported their results in 1961 at the Kyoto CR conference: there were clearly primary L nuclei, although not as many as had sometimes been suggested.

With the presence of primary L nuclei at the top of the atmosphere firmly established, there was steady improvement over the next few years in methods of particle identification, especially through the use of combinations of Čerenkov and scintillation detectors. Introduced in 1956 by Frank McDonald at the State University of Iowa, this combination has been widely used and has gradually clarified the charge spectrum through iron, the twenty-sixth element of the periodic table. Beyond this point, the situation was unclear for many years. On Earth the periodic table of the elements extends up to uranium ($Z = 92$), but the abundances drop steeply for atoms heavier than iron. If nuclei heavier than iron were among the cosmic rays, it was therefore anticipated that they would be rare. Some CR experiments had shown particles that seemed to be heavier than iron, but problems with their identification left us skeptical and the picture unsettled.

In the mid-1960s two quite different sets of observations demonstrated the existence of heavier particles among the cosmic rays. The first of these introduced a new technique to CR particle studies. Robert Fleischer, Buford Price, and Robert Walker at the General Electric Research and Development Center had been studying particle tracks formed in certain types of crystals and later revealed by chemical etching. Some clear plastics such as cellulose nitrate and Lexan were also found to register tracks of heavily ionizing particles. The GE researchers showed that measuring track dimensions could lead to particle identification, much as with tracks in photographic emulsions. The group's surprising discovery of heavy CR came from crystals extracted from meteorites, in which Fleischer, Price, and Walker found particle tracks that could only have been produced by particles much heavier than iron. Tracks such as these cannot be the products of radioactive inclusions in the meteorites, and the conclusion was inescapable—that the tracks had been produced by cosmic rays and had slowly accumulated throughout the meteorites' lifetimes, estimated to be in the hundreds of millions of years. The number of tracks found was in good agreement with what would be expected if the relative abundances of these heavy particles followed the generally known elemental abundances and if the CR intensity through those hundreds of millions of years had been about the same as today. The meteorite observations thus significantly extended the CR charge spectrum and also suggested that CR flow in the galaxy was not simply a recent phenomenon.

While the meteorites provided a record of heavy cosmic rays covering very long times, the discovery of heavy particles among today's cosmic rays came from photographic emulsions carried on a balloon flight in 1967. Peter Fowler of Bristol University increased the total collecting area of his emulsion package to more than four square meters and found many tracks definitely attributable to trans-iron nuclei, with one track possibly produced by a particle with a charge as large as 90, perhaps even uranium.

Through the next five years, balloons carried payloads of increasing size—flying barndoors, as we called them (Figure 5.6). The large areas that had to be scanned for tracks required a collaborative effort (an increasing trend in physics), initially involving Fowler, the GE researchers, and our group at Washington University. Complementing the emulsions were the new plastic detectors such as those made of Lexan; these had the advantage of not responding to lightly ionizing particles, which in a photographic emulsion would almost swamp the rare tracks of the heavy particles. A flight payload usually contained a sandwich of several layers of emulsion interleaved with plastics of different sensitivities. During a typical balloon flight 500 tracks of heavy particles would be found, about one in every 200 cm². Locating these tracks was a typical needle-in-the-haystack problem.

Once again, a technical advance yielded further significant improvements in the delineation of the charge spectrum for heavily charged particles. From an experiment conceived in 1969, it took ten years before a consortium of groups at Caltech, the University of Minnesota, and Washington University saw its 6 m² electronic detector launched on the third High Energy Astronomical Observatory (HEAO-3) satellite. A similar system was sent up by Fowler on the U.K. Ariel VI satellite. These systems were designed to examine the charge spectrum upward of about $Z = 20$; the HEAO satellite also carried detectors for a French-Danish collaboration to cover the lighter particles, with charges below 26.

Figure 5.6   *A barndoor system, ready for launch, at the National Scientific Balloon Facility in Palestine, Texas. Helium is entering the balloon through the two filling tubes. Ballast for control of altitude is contained in the package beneath the barndoor. Tiny Tim, the launch vehicle, was specially constructed for handling large loads. (Photograph courtesy of Herb Weitman, Washington University.)*

Not only did the HEAO-3 systems have large collecting areas and long exposure times, but they provided much better charge resolution than did the older emulsions and plastics. The HEAO charge spectrum is shown in Figure 5.7. To accommodate the wide range of relative abundances, the spectrum is shown in sections, with different scales. The rarity of the heaviest particles is clear: for every trillion (10^{12}) protons, fewer than two of any particle have Z greater than 60.

What we find is a spectrum that shows a persistent trend of diminishing abundances with increasing mass (charge), but with some important features superimposed. From carbon ($Z = 6$) through nickel ($Z = 28$), particles of even charge are more plentiful than adjacent particles with an odd Z. (This odd–even effect is well understood on the basis of nuclear properties.) No clear examples of particles with Z greater than 83 were found by HEAO, and the older reports of ultra-heavy particles, based on tracks in emulsions

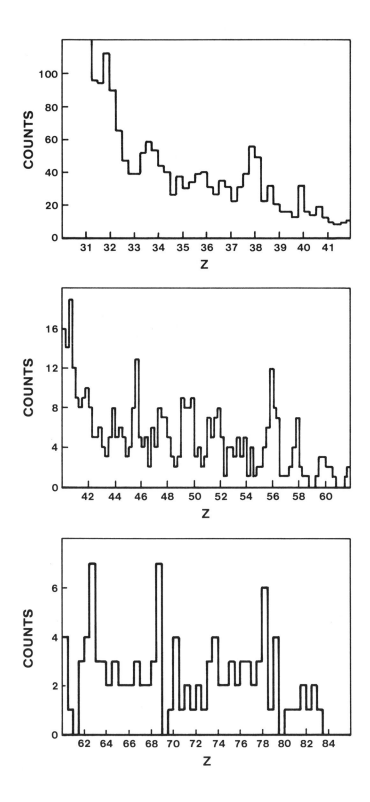

Figure 5.7
Charge spectrum for primary CR with Z greater than 31, as observed by the HEAO-3 satellite. The number of detected particles decreases steadily with increasing Z; no clear examples were obtained with Z greater than 83 (bismuth). (Based on data collected by the Caltech–Minnesota–Washington University group.)

and plastics, are now considered unreliable because of unsuspected calibration problems.

The best charge spectrum today comes from a combination of data from balloon and satellite experiments. Hydrogen and helium nuclei are so plentiful that adequate statistics have been accumulated during balloon flights, which have also produced excellent spectra up through iron. For the less abundant species and especially for the heaviest particles, the extended duration of satellite experiments is needed, but at present there is no immediate prospect of any significant improvement over present data. The size of the needed equipment and the long lead time necessary to design and construct a large experiment mean that any new major CR satellite is likely to be far in the future.

When we want to go beyond the elemental abundances and examine the proportions of individual isotopes, we can still make use of ionization measurements. Ionization energy loss depends on the particle's charge and speed but not on its mass. A particle can travel until its kinetic energy is exhausted; the distance this takes is called the *range*. Ranges can be computed from theory in terms of initial energy for particles of different mass. Different isotopes will have different ranges for the same initial energy. The relation between range and energy is the basis for our best measurements of particle masses and thus for discriminating between isotopes within a given element. Because important information is conveyed by the relative abundances of different isotopes, considerable technical ingenuity has been invested in designing experiments in which particle ranges are measured after initial measurements of ionization (energy loss rate). The range of a fast particle increases rapidly with its energy, so it becomes impracticable simply to build larger and larger detectors to stop higher-energy particles; after a while the required detector size would exceed the mfp, and too many particles would suffer collisions and fragment well before they could be brought to rest. The range method therefore works best with the less energetic particles, typically those with kinetic energy below about 300 MeV.

Several recent measurements of isotope abundances in cosmic rays have been obtained by groups at the University of California at Berkeley, the University of Chicago, and the University of New Hampshire. Among the beryllium (Be) nuclei, for example, 55 percent were found to be ^{7}Be and 39 percent ^{9}Be; radioactive ^{10}Be made up the remaining 6 percent. The precise proportion of ^{10}Be turns out to be of great importance in the theory of CR propagation. Carbon was found to be overwhelmingly ^{12}C; nitrogen had roughly equal amounts of ^{14}N and ^{15}N. Other isotope separations have been made, up through iron. Some of these isotopes were injected at the original sources while others have been produced by collisions en route to the earth, and the proportions carry interesting information regarding sources and propagation history.

The next category in the catalog of CR particles is electrons, once thought to constitute the majority of the cosmic rays but now known to amount to less than 2 percent of the number of protons. At first this might seem surprising, for if CR protons had started out from atomic hydrogen, there should be equal numbers of protons and electrons; thus their different abundances must be explained in the model for CR production.

Definitive detection of primary electrons was achieved in 1961 by James Earl of the University of Minnesota, using a cloud chamber, and by Peter Meyer and Rochus Vogt of the University of Chicago, using electronic detectors. High-altitude flights are required to avoid the many secondary electrons that emerge from interactions in the atmosphere. Since all electrons have identical masses, interest is confined to their energy spectrum and to the relative proportions of electrons with positive and negative electric charge—which introduces yet another new topic.

Atoms with their positively charged nuclei and negatively charged electrons constitute the matter that makes up our world. In principle, we could imagine another world in which nuclei of negative charge would be surrounded by positrons—a world of antimatter, made up of antiparticles. Antiatoms would have the same properties as familiar atoms; they would combine to form antimolecules and would even be the basis for antilife, physically no different from our own. From a distance, there is no way to distinguish between stars and antistars.

The theoretical foundation for antimatter was set out by P. A. M. Dirac in his quantum theory calculations in 1929. He suggested that particles and antiparticles have identical masses but opposite charges. A property of such mirror-image sets of particles is that they can be created in pairs (of particle and antiparticle) from energy. When they meet they annihilate, their electric charges neutralizing each other and their masses being converted back to energy. In both creation and annihilation, energy and mass convert into each other, via $E = mc^2$, as first suggested by Einstein. In electron-positron annihilation, the energy emerges in two or three X-ray photons, and in proton-antiproton annihilation the result is gamma rays or mesons.

This scenario is not simply a science-fiction product of the imagination. The predictions of Dirac's theory have been repeatedly confirmed. In 1932 antielectrons (positrons) were discovered in cosmic rays, and antiprotons were first seen in 1956 in an experiment at the new Bevatron accelerator at the University of California's Radiation Laboratory in Berkeley. Not unexpectedly, antiparticles enter into CR physics. Positrons are produced when sufficiently energetic photons (above 1 MeV) convert into electron-positron pairs; they also come from the decay of mesons produced in nuclear collisions. Much more energy is needed to produce a proton-antiproton pair, and as a result antiprotons are rare. The detection of antiparticles among the cosmic rays is thus a clue to physical processes that have occurred. We do expect some antiprotons to be produced by CR collisions with interstellar atoms. However, distinguishing between particle and antiparticle cosmic rays requires techniques that differ from the ones previously described, because particle ionization depends on Z^2 and we find the same energy loss rate whether the particle has a positive or a negative charge.

There are two ways to identify antiparticles. One method utilizes the bending property of a magnetic field that forces charged particles to follow curved paths. The degree of curvature depends directly on the field strength and the particle's charge, and inversely on the particle's mass and speed (its momentum). Because the charges can be positive or negative, the direction of curvature immediately shows the sign of the charge. The other method of identifying antiparticles rests on their property of annihilation in collisions with their particle counterparts. If the total energy of particles coming from a collision is much more than the incoming kinetic energy, then it is reasonable to assume that the added energy has come from annihilation.

The first observation of CR antiprotons came in 1979 in a balloon experiment carried out by a group at NASA's Johnson Space Center in Houston. The flux of antiprotons was reported to be around 5×10^{-4} times the corresponding proton flux. Russian and American groups have also used magnets with counters above and below to identify the antiprotons by their curved paths. The early measurements of the antiproton flux appear to be rather higher than expected from calculations for their production in interstellar collisions. More recent observations failed to find any antiprotons, and further measurements are clearly needed to clarify this puzzling finding.

When an electron and a positron are about to annihilate they may first form *positronium,* which is very much like a hydrogen atom but with the positron playing the role of the usual proton as the nucleus; Martin Deutsch of MIT was the first to observe it. Annihilation follows rapidly and produces the two or three characteristic X-ray photons that are easily detected. These characteristic photons have been observed by X-ray astronomers. The process of annihilation may sound like an unearthly phenomenon, but in fact it is now a routine diagnostic tool of radiologists, who inject into a patient radioactive tracer isotopes that emit positrons. The tracers spread through the body, and the annihilation photons from the

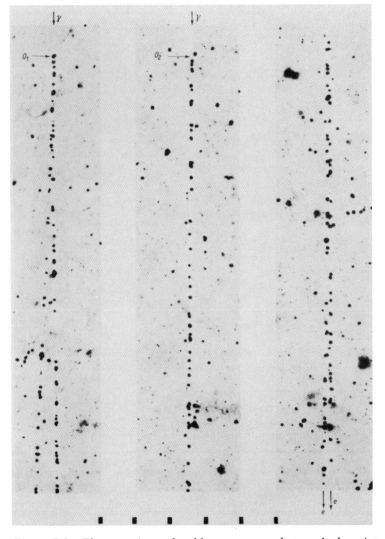

Figure 5.8   *Electron pairs produced by gamma ray photons. At the point where a gamma ray is converted into an electron-positron pair, the two particles are very close together. Their separate tracks can be seen after a short distance. (Photograph courtesy of Peter Fowler.)*

emitted positrons are then detected and mapped to produce images resembling X-ray photographs. This technique is called *positron emission tomography* (PET).

Annihilation was the signal observed by Meyer and Vogt in their discovery of positrons, not by direct detection of the immediate photons but rather by detection of the subsequent electron-photon cascade (Figure 5.8). The annihilation of antiprotons releases nearly 2,000 times more energy than for positron annihilation and provides the unmistakable signal that was observed by Andy Buffington and his colleagues (from Caltech and the University of California at Berkeley) when they were detecting CR antiprotons. The identification of heavier antiparticles such as antihelium is far less secure; no convincing demonstration has yet been made. Thus the existence of heavier antiparticles and the flux of the antiprotons remain unresolved questions. If the Space Station ever flies, it may carry the ambitious Astromag experiment with a powerful magnet for antiparticle identification.

In this chapter and the last I have cataloged the types of CR particles. The measurements that identify these particles have also been used to measure their energies, and I turn next to a discussion of the energy spectrum of the CR population.

The Energy Spectrum

The energy picture of the CR population is not a static one. All cosmic rays lose energy (generally only a little) as they travel through interstellar space, but some particles can have catastrophically destructive collisions with interstellar nuclei, fragmenting both target and projectile and creating mesons. Some cosmic rays leak out of the galaxy. If the overall population and energy content of cosmic rays are to remain at a steady level, then new particles must be injected to replace lost ones. Acceleration processes must maintain the energy spectrum. Once the energy gains and losses of cosmic rays are known, the energy budget can be determined, and only then can we hope to see which among the possible CR sources possess the necessary reservoirs of energy to inject new particles.

It will be helpful in this discussion to clarify the sometimes confusing sets of units used for measuring particle energies—a situation that has been made more complex by the fact that CR astrophysics brings together subfields that developed separately, each with its own jargon. Through the preceding chapters we have become familiar with the use of multiples of the electron volt, MeV and GeV, to measure kinetic energies of particles. For particles with much more energy, we simply use powers of ten. In addition, to avoid some cumbersome terminology, I will use eV and MeV from now on, with the understanding that eV/nucleon and MeV/nucleon are implied if the particle in question is a nucleus rather than simply a proton.

The electron volt and its multiples are employed for measuring the minuscule energies of single nuclear-sized particles. A larger scientific unit of energy, one that predates the electron volt, is the erg; the relation between erg and electron volt is expressed as $1 \, \text{eV} = 1.6 \times 10^{-12} \, \text{erg}$. (Forty billion ergs is the equivalent of one diet calorie.) Converting to ergs is a matter of convenience (and

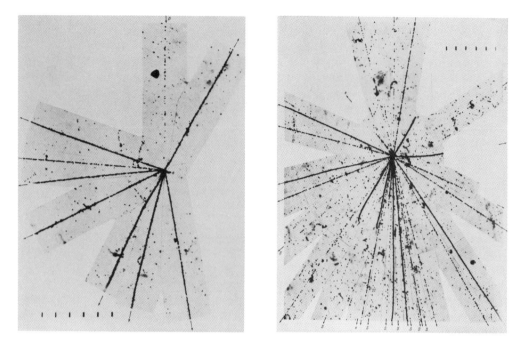

custom) when one is considering the total energy of a large number of particles.

To describe how rapidly energy is supplied or consumed the term "power" is used, which could be measured in eV/sec, but ergs/sec is preferred. When examining the CR energy budget, many researchers work in units of ergs/sec. Ten million ergs per second constitutes one watt (one horsepower is equivalent to 746 watts). A useful astronomical standard is the solar radiant power or luminosity, $L_\odot = 3.8 \times 10^{33}$ ergs/sec, equivalent to 3.8×10^{23} kilowatts. The rate at which solar energy reaches the Earth is about one kilowatt on each square meter.

Different methods must be used for measuring the energies of particles in different energy ranges. Ionization measurements for identifying particles often can also determine energies. For less energetic (slower) particles, those under about 300 MeV, the ranges are short enough that they can be measured in conjunction with ionization. To measure particles at intermediate energies, between about 300 MeV and 1,500 MeV (1.5 GeV), scintillation and Čerenkov counters can be employed.

For particles between 1.5 and 100 GeV, there are two current methods of measurement. One is based on ionization, which increases slowly for particles with energies above the relativistic minimum. Slow and fast particles can display the same energy loss, but the ambiguity is removed by using a Čerenkov detector, which identifies the fast particles. Use of the relativistic rise in

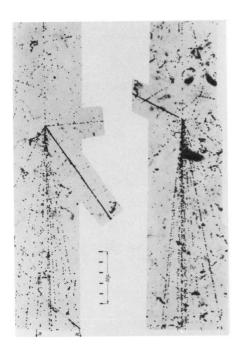

Figure 6.1
Nuclear disintegrations produced by protons of increasing energy. (left) 570 MeV: eight nuclear fragments each leave a heavily ionizing track, but no mesons are produced in this collision. (middle) 30 GeV: a heavy nucleus (silver or bromine) is torn apart by the incoming proton. Twenty-two heavily ionizing fragments emerge along with nine newly created "shower" particles, all pions. (right) 200 GeV and 300 GeV: so much energy is available that "jets" of mesons are produced. (Photographs courtesy of Peter Fowler.)

energy loss is generally confined to gaseous ionization detectors, since the increase in solid scintillators is too small to be useful. The other method in this energy region is the use of a magnetic spectrograph, which records the curvature of a particle's path in a strong magnetic field.

Up to around 100 GeV, all the methods of measurement are based on the signals generated along a particle's path. Above this energy, ionization and track curvature do not exhibit measurable differences with increasing energy; for example, scintillation detectors cannot distinguish between 500 and 5,000 GeV, and the magnetic fields available are not strong enough to produce measurable curvatures for the highest-energy particles. Consequently, less direct methods, based on nuclear collisions, are called for. This at once introduces a problem: we can measure the ionization signal of almost every particle that passes through a detector, but not all particles have collisions that can be analyzed. Many particles will not be so obliging as to have their collisions in detectors or in some other observable way; in the end, statistical corrections must compensate for the unobserved particles.

Which features of CR collisions can be exploited to assess the energy of the incoming particle? As collisions become more violent, the results gradually change in character, as seen in the "stars" in Figure 6.1. When an incident particle has only a few hundred MeV, some of this energy is transferred to the target nucleus, where it can promptly knock out a few particles. Some of the energy rattles around and is distributed, quite unevenly, among the remaining nucleons. Occasionally one of these nucleons accumulates enough energy to escape from the nucleus (it is said to *evaporate*), and, for example, the track of an emerging low-energy proton may appear. Perhaps a group of four nucleons may come out together as a heavily ionizing alpha particle, revealed by its dark track. We may even see short and thicker tracks of heavier fragments from the target. The characteristic time scale for evaporation is incredibly short by everyday standards, around 10^{-20} sec.

As the energy of an incident particle increases, it can generate mesons—mostly pions but also a few more exotic particles. Then we see the tracks of shower particles, thin tracks with well-separated grains, produced by mesons usually moving with speeds close to that of light. The number of mesons, the *multiplicity*, increases with primary energy, yet there are large fluctuations from one collision to another. At the same time the angular spread, the cone of angles encompassing the jet of mesons, decreases as the energy increases. The opening angle of this cone provides a measure of the energy: a $1/20°$ cone corresponds to about 1,000 GeV. The emission of mesons in these collisions takes place within about 10^{-23} sec of the initial impact, long before evaporation begins.

Charged mesons, which produce the tracks seen in emulsions and cloud chambers, are accompanied by neutral pions, which are not immediately visible. Like all mesons, neutral pions are unstable, but their half-life is extremely short. After around 10^{-16} sec, they usually decay into two high-energy gamma rays, or (in just over 1 percent of the cases) directly into an electron-positron pair and a photon. The gamma rays soon convert into more electron-positron pairs, and these in turn produce photons which convert into more pairs. The ensuing electron-positron-photon cascade builds up quite rapidly, then fades away as the energy is divided among too many particles for them to be able to reproduce further. In addition to the electron-photon cascade, another smaller cascade occurs that consists of protons, neutrons, and charged pions, all of which collide further and produce yet more mesons.

Because a cascade develops differently in the atmosphere and in solid detectors, methods of estimating primary energy are based on different ways of sampling showers to determine the growth pattern. At one time, energy estimates were derived by laborious microscopic analysis of hundreds of tracks in individual collisions recorded in photographic emulsions. These studies led to considerable understanding of high-energy collisions many years before particle accelerators could approach such energies. Today we use *calorimeters*, devices that measure simultaneously the total energies of many particles rather than measuring particle by particle. An ionization calorimeter consists of several ionization detectors separated by slabs of material in which the shower can develop. The detectors thus sample each cascade's growth; when this growth pattern is compared with the well-developed shower theory one can calculate the total energy for each event. As with other electronic detector systems, the calorimeter enables many more events to be handled automatically than with emulsions (no matter how picturesque that technique has been). The calorimeter technique is well suited to the energy range of 10^{12} to 10^{15} eV; at higher energies the cascade is too extended to be confined within a calorimeter of any convenient laboratory size. Fortunately, the Earth's atmosphere again comes to our rescue.

After a high-energy collision in the upper atmosphere, a cascade develops until (after only about 10^{-4} sec) a swath of millions or even billions of particles arrives at sea level, spread out over an area that can extend for several square miles. In these *extensive air showers* (EAS) a concentrated core of particles is arrayed around the shower axis, with fewer particles at larger distances. The sheet of particles is a few meters thick in the center and perhaps one hundred meters thick well away from the center. An array of counters, spread out in carefully arranged patterns, can detect only a small fraction of the shower particles but still enables researchers to reconstruct the shape and size of the shower and its energy. Thus, even though only a handful of the highest-energy particles arrive each year—far too few for any conventional detector to capture—their multiplication in the atmosphere gives them away.

The highest energy yet claimed for individual particles is just above 10^{20} eV. For comparison, the average kinetic energy of a molecule in water is in the region of 0.01 eV, and the total kinetic energy of all 3×10^{22} molecules in 1 cubic centimeter of water is roughly the same as that of a single CR particle with 10^{20} eV. If we could convert that one cosmic ray's kinetic energy to heat energy, it could boil the cubic centimeter of water, starting from just above freezing.

The discovery of air showers came accidentally in 1938 in France, where Pierre Auger and his colleagues were studying coincidences in a CR telescope and noted that they were recording coincidences between two counters that were several hundred meters apart. Further tests, some by Auger and others by George Rochester and J. G. Wilson at Manchester University using cloud chambers as well as counters, showed that large numbers of particles were arriving simultaneously, within the timing capabilities of their circuits.

Studying an EAS and identifying its various components requires different detectors, some with lead shielding to filter out the soft component (the electrons and photons). Scintillators and Čerenkov counters are in widespread use, in arrays extending over

Figure 6.2   *General view of the Fly's Eye detector on top of Little Granite Mountain, Utah. The array consists of 67 units, each with a 62-inch mirror and phototubes that can be pointed in different directions at night, to collect the faint flashes that signal the arrival of high-energy CR showers. (Photograph courtesy of G. L. Cassiday, University of Utah.)*

distances of many kilometers. The best EAS data in recent years have come from systems near Leeds, England; Volcano Ranch, New Mexico; Salt Lake City, Utah; Sydney, Australia; and Yakutsk, in Siberia. A Chinese-Japanese collaboration also has an array of detectors.

The group at the University of Utah has adopted a novel approach with their Fly's Eye Detector (Figures 6.2 and 6.3). Whereas the usual EAS system is designed to detect shower particles arriving at ground level, the Fly's Eye consists of an array of mirrors and sensitive phototubes that collect the faint light produced by the EAS particles on their way down through the atmosphere. The Utah group seeks out light emitted by atmospheric

Figure 6.3 *Close-up view of one of the Fly's Eye units, showing an array of fourteen light-detecting phototubes at the mirror focus. The housing, just over 2 m across, is motorized and can be driven to point in desired directions. During the day each unit is covered and aimed down, to avoid damaging the sensitive phototubes by sunlight. (Photograph courtesy of G. L. Cassiday,* **Nuclear Instruments and Methods** *240 [1985]: 410, reproduced by permission of North-Holland Physics Publishing Co.)*

nitrogen atoms that have been hit by shower particles. They have compared their problem to that of detecting "a blue 5-W light bulb streaking through the sky at the speed of light against a continuous backdrop of starlight, atmospheric airglow, and manmade light pollution. In addition, sporadic sources of lights such as lightning, auroras, airplane, and smokestack strobe lights . . . create a certain visual havoc" (*Nuclear Instruments and Methods,* p. 412).

In an EAS, the charged pions that do not collide will decay (in about two hundred millionths of a second) into charged muons, many of which reach detectors. Energetic muons constitute a durable component in the EAS; they behave a lot like electrons but are two hundred times heavier. Unlike nucleons and charged pions, muons rarely interact with nuclei; they are forty thousand times less likely to produce the photons needed for shower growth. Consequently, the only important way a muon can lose energy in air is by gradually transferring small amounts of energy to electrons in the atmospheric atoms. Muons can thus arrive at ground level with much of their original high energies intact and then penetrate deep inside the Earth. Muon detecting systems have been operated down to depths of 12,000 feet below the Earth's surface in gold mines near Johannesburg in South Africa and in the Kolar gold field in India. Another deep counter array is located in a tunnel under Mont Blanc in the Alps. Primary protons with energies of around 10^{16} eV produce the muons that are seen in these counters; no other charged particles can penetrate to such great depths.

Because the shower's primary particle is not observed directly, its energy and nature must be deduced from the products of collision, which multiply and then are detected. The precise mix of muons, electrons, and nucleons and their varying proportions across the shower front provide clues to the nature of the primary particle. The data are ambiguous, but from present evidence we think that the dominance of protons is preserved to the highest energies. There is some indication that the heavier primaries (such as the C, N, O group and Fe) are in their usual abundances, at least up to 10^{16} eV (Figure 6.4).

A rather different identification has been entertained for some of the highest-energy primaries: high-speed dust grains. Huge amounts of microscopic dust particles pervade interstellar regions like a celestial smog, absorbing light preferentially at the shorter (blue) wavelengths and making distant stars appear more red. A few of these grains, it has been suggested, might be accelerated and could, on colliding with atoms in the atmosphere, produce an EAS. Although this hypothesis is sometimes raised as a possible explanation for the highest-energy EAS primaries, the best identification still seems to be the conventional protons and a few heavier nuclei.

EAS can also be produced by gamma ray photons with sufficiently high energy; these are identified—with probability but not certainty—from the resulting shower structure. Photon-initiated showers start with the primary photon converting into an electron-positron pair, and then the usual electron–photon cascade follows. If the primary projectile is a photon rather than a nucleon, the shower will contain very few pions, and although high-energy photons can fragment a nucleus, this is rare, so the muon and nucleonic components will be much reduced from their normal proportions. Showers that are muon-poor are likely candidates for

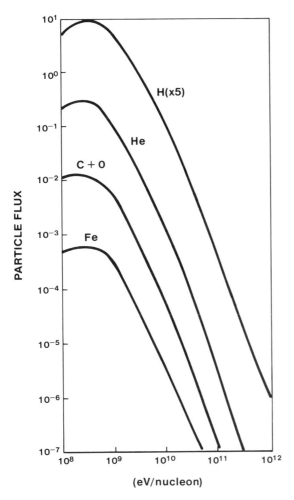

Figure 6.4 *Energy spectra for some of the most abundant CR particles. The hydrogen spectrum has been raised to distinguish it from the helium line, while carbon and oxygen have been combined. The general shapes of the spectra are similar, though there are definite differences in slope in the high-energy region. Spectra below 100 MeV/n (not shown) are strongly influenced by the solar wind.*

having been produced by photons, but the problem is that the muons might simply have missed the detectors rather than being truly absent.

Although photon primaries seem to be rare or absent among most regular EAS that come randomly from all directions, there is growing evidence that some showers are produced by photons coming from a very few astronomical objects such as Her X-1 and Cyg X-1. These objects, located in the constellations Hercules and

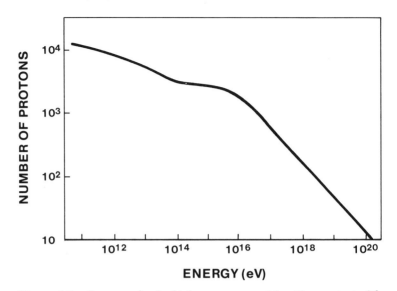

Figure 6.5 *Spectrum for the highest-energy particles. To accentuate differences in spectral form, the data points have been adjusted so that a spectrum $E^{-2.5}$ would appear as a straight horizontal line. The persistent slope of the actual line indicates variations in the energy dependence, generally steeper than the value 2.5.*

Cygnus, are already known as emitters of X-rays and gamma rays, and the axes of their associated showers point directly to them as they move across the sky. Photons of around 10^{12} eV have been identified with Cyg X-1, whose behavior at different wavelengths is quite irregular. This object is thought to be one of the better candidates for a black hole.

Finally, we do not expect to see many EAS initiated by electrons. An electron with enough energy to produce a shower probably will not reach the Earth; it is far more probable that it will have been degraded through collisions with the many interstellar photons. This effectively places a limit on the maximum electron energy that we observe, a subject I return to in Chapter 9.

When all the data are put together, we find that there is a flux of about 1 particle/cm^2 sec at 10 GeV. Above 10^{20} eV, we can expect to see only about 5 particles *per century* per square kilometer! This is an intimidating prospect, but the shape of the top end of the energy spectrum is sufficiently important that the infrequent arrival of these particles is being monitored by several EAS groups.

Regarding the shape of the energy spectrum, we find, as with the solar CR, that the number of particles decreases with increasing energy, proportionally as E^{-a} (Fig 6.5). To a fair approximation, there is little (but important) variation in a around the value 2.6, from 10^{10} eV on up to around 10^{15} eV. Then the spectrum steepens and a has a value closer to 3 until 10^{19} eV, where there is a suggestion of a smaller value. The importance of these changes in a lies in their relation to the way in which the cosmic rays travel in the galaxy and to the physical processes in which they can be accelerated, lose energy, or even escape from the galaxy, as will be described in Chapter 9.

Below 10 GeV the situation is just as complex as with higher energies, but there are differences. The energy spectrum bends over and displays a broad maximum around several hundred MeV, then decreases before beginning to rise again at the lowest energies observed, around 10 MeV. In addition, the low-energy spectrum varies significantly over the 11-year solar cycle, most noticeably at the lowest energies. This *modulation* of the energy spectrum is understood in only the most general of terms. Arriving cosmic rays have to traverse the solar wind as it blows out past the Earth, and they lose energy. A typical energy loss can be 400–600 MeV, so a particle detected with 1,000 MeV would originally have been in the range 1,400–1,600 MeV before it encountered the solar wind. The residual energy is clearly related to the solar wind's strength at that time. We may simply not be able to detect any particles whose distant energies are well below 400 MeV. For this reason our knowledge of the unmodified shape of the energy spectrum in interstellar space is shaky below 1 GeV, yet this is potentially the energy region with most of the cosmic rays.

When we turn our attention to the energy spectrum of the heavy CR particles, those that are not protons, we find that the E^{-a} shape still provides a good framework for analysis. The energy spectrum of the helium nuclei is similar to that of the protons and so, in general terms, are the spectra of the other heavy particles, which can be grouped into three categories. The first includes nuclei such as C, O, Si, and Fe, which seem to be plentiful in CR sources and whose intensities near the Earth represent mainly nuclei that have survived their long journeys. The second group of nuclei, such as Li, Be, and B, are virtually absent in the source regions and are produced when particles from the first group fragment. Finally, the third group consists of particles that are mixtures of primaries and secondaries. Examining differences in the relative proportions and energy spectra of these three groups tells us a great deal about the life cycle of cosmic rays.

These measurements of CR energy can be summarized in several ways. We have seen in Figure 6.4 a graphic display of the numbers of particles with different energies, and a selection of those data are set out in Table 6.1, in which the numbers of particles

Table 6.1   *Average fluxes of primary cosmic rays at the top of the atmosphere*

Type of nucleus	Flux (particles/m² sec)
Hydrogen	640
Helium	94
Lithium, beryllium, boron	1.5
Carbon, nitrogen, oxygen	6
Nuclei with $Z = 9$–25	1.9
Iron ($Z = 26$)	0.24
Cobalt, nickel ($Z = 27, 28$)	0.01
All nuclei with $Z > 28$	0.003

Note: All flux values refer to particles with energies above 1.5 GeV/n, arriving at the top of the atmosphere from directions within 30° of the vertical.

are listed that would be registered each second in a 1 m² detector flown above the atmosphere. If we add up all the energy carried by all of the CR particles, we find that the rate of arrival of CR energy on the Earth amounts to about 100,000 kilowatts (10^5 kw)—about one billion times less than the energy arriving in sunlight, but comparable to the total energy that we receive in starlight.

In the following chapters I explore the implications of these numbers as I attempt to disentangle the travel histories of the cosmic rays and consider their possible sources.

Nuclear Clues

*T*racking down the origin of cosmic rays is a task that has followed many twists and turns and has not yet been completed. We have accumulated a great deal of information about many aspects of cosmic rays. We know the major characteristics of the solar CR and the solar wind's effects on the galactic CR. The relative abundances of nuclei have been well charted, in detail up to atomic numbers around 60, and major features have been determined for the rest of the periodic table. We have made a good start at identifying individual isotopes. The energy spectra of the more abundant species have been measured for the energy range that encompasses more than 95 percent of the total CR energy. How can all this information be pulled together so that we can start to construct a coherent picture of cosmic rays?

In a popular view of science, further progress comes through induction: we assemble an array of well-established specific facts and from them, by compelling logic, construct a general theory, broad in its sweep. Some scientific problems readily yield to this form of attack, but many, perhaps most, do not. Instead, a more fruitful approach is what Karl Popper has termed conjecture and refutation. In this widely followed procedure, a conjecture (or model) is suggested and its consequences explored by logic, deduction, and computation. The confrontation between predicted consequences and actual observations can either lend support to the model or can require its modification or even rejection. Where does the conjecture or model come from? There is no simple or unique answer. Models come from hunches, flashes of inspiration, usually based on experience or on comparison with problems that seem similar. They emerge after the exhaustion of alternatives. "They can be reached only by intuition, and this intuition is based on an intellectual love of the objects of experience," as Einstein once said (*Imagination and the Growth of Science,* p. 9).

The model for most cosmic rays that is now generally accepted—at least as a starting point—locates their origin and confinement inside our own Milky Way galaxy. The initial charge and energy spectra of the primary particles are assumed, and their subsequent propagation is studied. The particle energies are tracked through any further acceleration or energy losses. Some cosmic rays escape from the galaxy; others have collisions that produce secondary fragments. Changes in the relative abundances of different particles are tracked through collisions that remove some of the primaries and create secondaries such as the L nuclei. The energy spectra and relative abundances are part of the input, along with the various mean free paths and cross sections.

Finally, calculations yield predictions for what we can expect to observe at Earth. These computations include many physical quantities (parameters); some of their values can be measured in the laboratory while others must be estimated and later adjusted to agree with observations. The more adjustable parameters a model contains, the closer we can come to "explaining" our observations; on the other hand, we might have less confidence in such a model because different sets of parameter values might yield equally good fits. A good theory keeps its adjustable parameters to a minimum.

The composition or relative abundances of cosmic rays clearly provide a logical starting point for model building. We can start by comparing the relative abundances of CR nuclei to abundances in the solar system, the material most accessible to us. We find a broad similarity and a striking difference. The similarity shows in a steady trend toward lower abundances as we proceed from the lightest nuclei, hydrogen and helium, to the heaviest. The immediately obvious difference lies in the abundance of the L nuclei (lithium, beryllium, and boron), which are quite plentiful in the CR beam but about 100,000 times less abundant in the solar system. What does this tell us, and why are there so few L nuclei at large? The answer comes from one of the great success stories in astrophysics, the theory of nucleosynthesis.

Our present best model for the evolution of the universe starts with an extraordinary explosion, the "big bang," when all the contents of the universe were jammed into an unimaginably small volume. In those first few high-temperature minutes, most of the helium in the universe was created by hydrogen fusion. Although helium is still being formed inside stars today, this source is not sufficient to produce the known amount of helium. Recognition of the cosmological production of helium during the big bang solved this long-standing problem. Deuterium was also produced in the big bang, and the proportions of deuterium and helium today provide clues to the conditions then, about 15 billion years ago. Expansion of the universe, accompanied by a drop in the temperature, later permitted protons to combine with electrons to form hydrogen atoms.

Much later the dispersing material began to clump into galaxies, and after some hundred million years the first stars began to form, composed mostly of hydrogen. The time scale for further development depends on the mass of each star, but the initial behavior is qualitatively similar for all stars. The gravitational attraction between the atoms that make a star produces a high pressure at the star's center. Hydrogen subjected to such a high pressure will be at a very high temperature. In the sun this central temperature is about 15 million degrees; once again the hydrogen is totally ionized into protons and electrons. At such high temperatures, colliding protons can fuse to create deuterons. Further collisions and a web of nuclear reactions lead to the production of helium nuclei, with the release of considerable quantities of energy. This energy, much of it in the form of high-energy photons, percolates slowly to the surface from which it is radiated outward, some of it arriving on Earth to keep us warm.

In each chain of fusion reactions, four hyrogen nuclei combine to form a single helium nucleus. The total number of nuclei is therefore reduced—one helium nucleus exists where before there were four protons, and a helium nucleus is not much larger than a proton. With this reduction in volume, the star's central core cannot withstand the pressure of the overlying material and contracts, raising its central temperature. When the central temperature reaches one hundred million degrees, helium nuclei fuse in further reactions and carbon and oxygen nuclei begin to form (out of three and four helium nuclei, respectively). For a star of solar size, this stage is reached after around 10 billion years and is accompanied by a general expansion, into the red giant phase.

For more massive stars, the pace of evolution quickens and the formation of nuclei can proceed beyond oxygen. Some reactions release neutrons, and these can be captured by other nuclei. In many stars the buildup of heavier nuclei occurs by this steady accumulation of neutrons. Some new nuclei are radioactive and decay before another neutron can be captured, but the accretion continues through successive neutron captures. This slow (or *s*) process ultimately produces nuclei as large as lead ($Z = 82$) and bismuth ($Z = 83$), the two heaviest nonradioactive nuclei.

An interesting feature of these nuclear reactions is that L nuclei are created but just as rapidly are consumed in further nuclear reactions. The general rarity of these three L elements (lithium, beryllium, and boron) can now be seen as a natural consequence of nuclear properties and the reactions taking place in the interiors of stars. In contrast, the only viable theory for the presence of the relatively large numbers of L nuclei among cosmic rays is the fragmentation that some of the heavier CR particles undergo during interstellar travel. The L nuclei are thus important as indicators of the distances traveled by cosmic rays and the resulting numbers of collisions. Moreover, it appears that almost all of the Earth's content of these three light elements was produced by interstellar CR collisions.

In some massive stars the cycle of central fuel consumption, subsequent collapse, and further heating becomes unstable, and as a result many more neutrons are generated during an explosive stage. The torrent of neutrons (in this rapid or *r* process) can be so large that many neutrons are captured and nuclei heavier than lead or bismuth are produced before radioactive decay shuts off this growth. These heavy nuclei are all radioactive, and most have short half-lives. Only thorium ($Z = 90$) and uranium ($Z = 92$) have isotopes with long enough half-lives that they have survived the four and a half billion years of the solar system's existence. Their presence on Earth proves that the solar system formed from interstellar material that had already been enriched through the *r* process. Short-lived nuclei with atomic numbers between 84 and 89 and beyond 93 do not occur in nature, but during the past forty years they have been synthesized in large accelerators and also in test nuclear explosions.

The explosive *r*-process stage of a star's history occurs during an event that appears as a supernova—the extreme, sudden, and unpredictable brightening of a star, as happened so spectacularly in the Large Magellanic Cloud in February 1987. In such an explosion, newly created heavy nuclei are hurled out into the interstellar medium. Some drift into regions of higher density, where they are among the atoms condensing to form the next generation of stars.

The theory of hydrogen fusion reactions in the sun is based largely on the pioneering work of Hans Bethe of Cornell University, who won the Nobel Prize in 1967. For heavier nuclei, the classic research was carried out by Geoffrey Burbidge, Margaret Burbidge, Willy Fowler, and Fred Hoyle of Caltech and Cambridge University. Their 1957 paper (usually referred to as B^2FH) set out the roles of the *r* and *s* process. Quite independently, Al Cameron, in Canada, derived similar results at the same time. Of course, the theories of stellar evolution and nucleosynthesis are far more complex than the brief summary I have given here, and other processes in addition to the *r* and *s* come into play.

The properties of different nuclei and their behavior in nuclear reactions can often be studied in the laboratory, but sometimes formulas have had to bridge the gaps. With computers, one can model a stellar interior or a star's evolution during the catastrophic supernova stage. Some of this analysis can be used when observed abundances are tabulated, element by element and isotope by isotope, by looking at materials from the Earth, moon, and meteorites as well as data from astronomical spectral analysis. Solar system material does not come from only the *r* or the *s* process but appears to consist of a mixture, with additions from other nucleosynthetic reactions.

Another outcome of the *r*-process calculation is a prediction of the abundances of the long-lived isotopes ^{232}Th, ^{235}U, and ^{238}U. Thorium has a half-life of 14 billion years, and after ten successive but unevenly spaced radioactive decays it creates a stable isotope of lead, ^{208}Pb. The uranium isotopes produce ^{207}Pb and ^{206}Pb. Knowledge of the half-lives and abundances of all these isotopes has enabled researchers to estimate the age of the solar system at 4.6 billion years. This result provides a calibration for the time scale for stellar evolution, and from it we can infer the age of our galaxy to be 10 to 20 billion years.

The importance of these studies for a cosmic ray model is that there may be signatures in the CR charge spectrum that tell us about their sources. If, for example, cosmic rays come from supernovas with their *r* process, then we would expect to detect some transuranic nuclei. Other clues to the roles of the *r* and *s* processes, in addition to the presence or absence of the heaviest nuclei, are the proportions of intermediate nuclei among cosmic rays.

The approach to CR modeling should now be coming into clearer focus. We can assume that CR particles start from a certain type of stellar source, and we can further assume a set of relative abundances (such as from *r* or *s* processes) at origin. We can then trace the history of these particles as they travel through the interstellar medium, and the L nuclei can be used as the first distance indicator. In carrying out these calculations, we must allow for the ways in which fragmentation and energy loss rate depend on particle energy.

With the background just described, we are ready to assemble and assess a comprehensive model for cosmic radiation. After allowing for effects of the geomagnetic field, we find that, in general, galactic CR arrive at the top of the Earth's atmosphere uniformly from all directions. This isotropy can mean either that the particles do not come from any one region or that they have been deflected so often in their travels that they have lost all sense of their original directions of travel. Two pieces of astronomical evidence are relevant to the puzzle.

Most cosmic rays do not come from the sun, so it is plausible to look for sources of galactic CR among stars that are more luminous than the sun or more energetic in some other way. But stars are definitely not distributed at random; rather, they are concentrated in the equatorial region of the galaxy, in the Milky

Way. If cosmic rays traveled along straight lines from stellar sources to Earth, we should see more of them arriving from the galactic disk, but that is not what we observe. Thus we must look for alternative explanations for the isotropy.

Across the enormous interstellar distances there is a general magnetic field, well organized in some regions and more tangled in others. The strength of this field, deduced from astronomical measurements, is about 10^5 times weaker than the magnetic field on the surface of the Earth. On the average, the energy density of this magnetic field is around 1 eV/cm^3, comparable to the energy density of starlight. Another indication of the presence of the interstellar magnetic field comes from radio astronomy. Radio signals have been detected from many regions that are remote from clearly identified objects such as stars and galaxies. The spectra of these radio signals unambiguously identify their origin in *synchrotron radiation*, in which high-speed electrons travel along curved paths in a magnetic field and emit radio waves. This cosmic radio noise (as it is sometimes called) leads to an estimate of the magnetic field strength that agrees well with that obtained from optical measurements. Another measurement of the interstellar magnetic field comes from pulsar radio signals, and these quite different methods yield general agreement.

Because the cosmic rays are electrically charged particles, they are forced to travel along curved paths in the interstellar magnetic field. In places where the field's irregularities are moving, the cosmic rays may be accelerated or slowed down in addition to being deflected. In general, we expect the CR paths to be severely tangled.

When a charged particle in a magnetic field moves around a circular arc, the radius of that circle depends on the field strength, B (measured in microgauss), as well as on the particle's energy, E (in GeV) and charge, Z (in units of the fundamental charge on the electron). For high-energy particles, the formula that expresses this relation (in units that are most appropriate for CR and interstellar conditions) is:

$$R = 3.3 \times 10^{12} E/(ZB),$$

where the radius is in centimeters. We encountered this behavior of charged particles in Chapter 3 when examining the effects of the Earth's magnetic field. On the galactic scale the field strength is much weaker, but the size of the region in which particles travel is far larger than the solar system. Table 7.1 lists the radii of curvature of the paths of CR particles having typical energies for an interstellar field of 3 microgauss.

These numbers tell us that, for example, a 1–GeV proton will be moving along a circular arc of radius 1.1×10^{12} cm in a typical

Table 7.1 *Radii of curvature of proton paths in 3-microgauss field*

Proton energy (eV)	Radius (cm)
10^9	1.1×10^{12}
10^{11}	1.1×10^{14}
10^{15}	1.1×10^{18}
10^{17}	1.1×10^{20}
10^{19}	1.1×10^{22}

interstellar field. This dimension is less than one-tenth of the Earth-sun distance, so the 1–GeV particle will have its path coiled up on a scale that is small compared to the solar system and minuscule in the galaxy. If the magnetic field extends over a region much larger than that radius, then a CR proton can be confined within the region. In contrast, if the extent of the magnetic field is much smaller than that radius, the proton will not be confined, even though for a short while it will follow that circular path. At the high-energy extreme, the radius of curvature for a particle of 10^{19} eV is comparable to the diameter of our galaxy and far larger than the thickness of the galactic disk. Thus the EAS primaries of such high energies cannot be effectively contained within our galaxy, and this immediately introduces the possibility of an extragalactic origin, at least for the highest-energy particles—a subject I return to in Chapter 8.

Because the paths of cosmic rays are so tangled, the important thing to know is the actual distance traveled. This is like measuring the length of a piece of string after straightening it out rather than using the overall size of a big knot. As cosmic rays travel, they occasionally encounter atoms in the interstellar medium, and the number of collisions (with resulting fragmentation or creation of mesons) is proportional to the total number of atoms along the path (Figure 7.1). If the interstellar medium had the same density (number of atoms per cubic centimeter) everywhere, we could simply convert the path length in centimeters to the equivalent number of atoms encountered. Where the density is irregular, we can still figure out how many atoms are along the path if we allow for the variations in density.

To introduce a familiar analogy, the behavior of CR particles could be compared to the experience of running along the concourse in an airport. The distance you actually run will certainly be longer than the direct distance to the gate, and the number of your near collisions will depend on the density (numbers) of passengers, as well as their sizes and speeds. You will have fewer collisions when you run with the stream of other departing passengers than

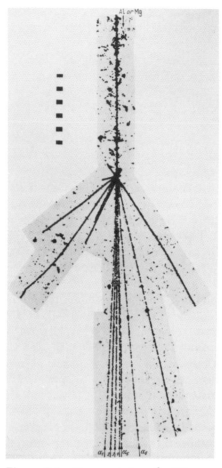

Figure 7.1   *Fragmentation of a CR magnesium nucleus (charge Z = 12) into six high-speed alpha particles which emerge in the tight bundle of tracks. The more widely spread black tracks are fragments of the target nucleus in the emulsion. Collisions of this sort, between primary CR particles and interstellar atoms, provide a means by which we can estimate the distance and time of travel of the cosmic rays. They may also be the major source of lithium, beryllium, and boron found on Earth. (Photograph courtesy of Peter Fowler.)*

against a stream of those arriving. The total distance you run must be measured along your zigzag path, or it could be expressed in terms of the number of near-collisions.

   With CR travel it is customary, and more useful physically, to express distances by the total mass of all atoms encountered, and to do so in units of grams per square centimeter. Imagine the path straightened out (Figure 7.2) and a particle traveling a distance D cm through a region containing n atoms/cm^3, with each atom

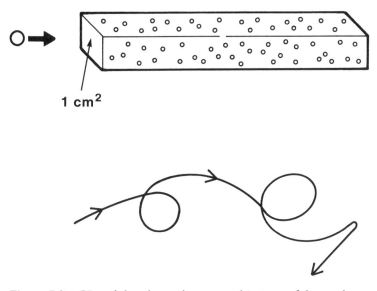

1 cm²

Figure 7.2 *CR path lengths can be expressed in terms of the number or total mass of atoms encountered. Along a straight path, the number of atoms within an area of 1 cm² provides this measure, usually given in g/cm². Along an actual path, much bent by the interstellar magnetic field, the equivalent straight length is obtained by adding up the contributions from many small and relatively straight segments.*

having a mass m grams. Then a total of nmD grams (of interstellar matter) will be within a volume that is D cm long and 1 cm² in cross section. For example, in the interstellar medium, a typical density is 1 hydrogen atom/cm³, and each atom has a mass of 1.6×10^{-24} gram. A CR path of 10^{21} cm can then be expressed as

$$(1 \text{ atom/cm}^3) \times (1.6 \times 10^{-24} \text{ g/atom}) \times (10^{21} \text{ cm})$$
$$= 1.6 \times 10^{-3} \text{ g/cm}^2.$$

The diameter of the Milky Way galaxy is 10^{23} cm, equivalent to 0.16 g/cm² if the density everywhere were 1 atom/cm³. How many collisions a cosmic ray actually undergoes also depends on the sizes (cross-sectional areas) of the cosmic-ray and the interstellar atoms, but the g/cm² figure provides a general index to CR travel history.

These calculations can also be applied to the propagation of carbon and oxygen nuclei, which are plentiful in prospective CR sources in contrast to the L nuclei—Li, Be, and B—which are very rare. The L nuclei that we detect among the cosmic rays at Earth must be fragmentation products from interstellar C and O collisions. We can calculate that C and O nuclei must have traveled along paths of around 5 g/cm² to produce the observed numbers of

L nuclei. The distance is somewhat dependent on C and O particle energies, being about 7 g/cm^2 for 1-GeV particles and 2 g/cm^2 for 10-GeV particles. For an average density of 1 atom/cm^3 in the interstellar medium, the 5 g/cm^2 path corresponds to a distance of $5/(1.6 \times 10^{-24}) = 3 \times 10^{24}$ cm, or thirty times the galactic diameter. The CR paths must indeed be tangled; from the L-nuclei abundance we deduce that a typical C or O particle has traveled a distance of about thirty times the galactic diameter between its origin and Earth.

We can also calculate the time taken for this long journey. For a particle of 1 GeV, the speed is close to nine-tenths the speed of light, or 2.7×10^{10} cm/sec. The 5 g/cm^2 will be covered in 3×10^{24} cm/$2.7 \times 10^{10} = 10^{14}$ sec, or about 3 million years. How reliable is this estimate of the travel time? It is based on distance (expressed in g/cm^2), speed, and interstellar density. The g/cm^2 value comes from the abundance of the L nuclei (the values are slightly different when one is using iron nuclei and their corresponding fragmentation products), and this figure is probably good to within about 50 percent. The speed is well known, but the value assumed for density (1 atom/cm^3) is at best a guide. Since density varies greatly across the galaxy, it is necessary to introduce a calculation quite different from any we have seen thus far to determine the density where cosmic rays travel.

Most nuclei assumed to be injected at CR sources are stable (nonradioactive). Among the fragmentation products generated during their subsequent travel, however, there are many radioactive nuclei. If we start with an assembly of 1,000 radioactive atoms, then after one half-life, one-half of the original nuclei will remain and the others will have decayed into lighter nuclei, after emitting alpha particles or electrons. After another half-life 50 percent of the remaining nuclei will have decayed, so that only 25 percent have survived; after another half-life the survivors are down to 12.5 percent, and so on. The decays of radioactive nuclei start immediately after creation, and there is no way we can identify which nuclei will decay promptly and which will hold out for many half-lives. All we can do is describe the process in statistical terms, which turns out to be quite accurate.

Among the beryllium isotopes produced by the fragmentation of heavier CR nuclei, only ^{9}Be is stable, but ^{10}Be has a half-life of 1.6 million years—comparable to the travel time of cosmic rays. If the travel time were much longer than 1.6 million years, then most of the ^{10}Be nuclei would have time to decay, and we would not detect many among the CR particles. Conversely, if the travel time were much shorter, we would expect most ^{10}Be particles to survive and to show up when the Be isotopes were identified. Our best measurements show that 6 percent of the beryllium nuclei are ^{10}Be.

From this and from our knowledge of how C and O nuclei fragment, we can figure that their average travel time (for particles around 1 GeV) is close to ten million years. A similar calculation can be performed for an isotope of aluminum, ^{26}Al, which has a half-life of 700,000 years and is a fragmentation product of silicon and calcium. Then, if we invert the calculation and use these estimates of the travel time, we can refine the value for the average density of regions where cosmic rays travel to 0.2–0.4 atoms/cm^3.

Another product of these calculations is a set of abundances for cosmic rays at their sources, before interstellar fragmentation has begun. For example, the relative source abundances of carbon : nitrogen : oxygen of 0.8 : 0.1 : 1 give the best fit to the ratios observed near Earth (1 : 0.25 : 0.9). Similar analyses have been carried through for other nuclei, such as those with charges 21–25, many of which are fragmentation products from primary iron nuclei ($Z = 26$).

I should add a word of explanation here. The calculation just outlined is typical of many in astrophysics. For many processes, the relevant quantities cannot be pinned down with the precision expected under laboratory conditions; agreement to within 10 or 20 percent is often considered good (or fortuitous). Knowing a quantity to within a factor of two sometimes suffices to exclude some models and concentrate the attention of researchers on others.

Calculating the effects of propagation requires extensive computer codes, starting with an equation for each type of particle that represents all relevant quantities and processes. The gradual change in composition of a CR beam is tracked by computing the cumulative effect of successive changes during small increases in the distance traveled. Allowance is built in for collisions, fragmentation, energy losses by ionization and radiation, the creation or injection of some particles, and the decay of others. A set of interlocking equations is needed because fragmentation that destroys a nucleus of one type (such as C or O) will produce nuclei of other types (such as Li, Be, or B).

By now we know with reasonable precision the values for most terms in these equations; some terms may be negligible in special cases, such as when particles with energies over 10 GeV are considered. Moreover, we can investigate what happens when starting assumptions are changed; for example, we can assume different starting proportions of C, N, and O and examine the resulting production of L nuclei and ^{10}Be to obtain the best agreement with our measurements.

In carrying through these calculations, we must also make assumptions regarding the regions in which cosmic rays are moving. Is the travel entirely in the galactic disk, or is some part of it in the less dense halo? The currently popular model is that of a "leaky box," in which cosmic rays are mostly confined to the disk by the galactic (interstellar) magnetic field. Because the energy density of cosmic rays is comparable to that in the magnetic field, they exert a major influence on the field, which bulges out of the disk into the halo. Leakage of cosmic rays into the halo depends on their energy, high-energy particles have curved paths with larger radii of curvature and thus are more likely to get into the halo than less energetic particles with more tightly coiled paths.

The distances that particles travel depend on their energies. We see this when we examine the energy dependence of the abundances of fragmentation products such as the L nuclei. An extensive analysis shows that any given energy has a distribution of path lengths. Some particles travel considerably greater distances than others, and it has turned out that among heavier particles with high energies there is a shortage of short path lengths. Shorter path lengths produce fewer collisions and thus affect the production of secondaries.

In much of this chapter I have concentrated on nuclei in cosmic radiation and how their abundances reveal their propagation histories. I have neglected CR electrons, whose near-Earth presence can be observed directly and whose distant presence is detected by the radiation they emit. This is an important topic that I will return to in Chapter 9, but first there is more to say about the nuclear component and its origins.

The Origin of Cosmic Rays

Over the years, as our fund of information has grown, various models of the origin of cosmic rays have been entertained, supported, and sometimes neglected—or recycled. No model yet satisfactorily encompasses all of our knowledge. The changing views of partisans in this long-running debate provide a fascinating insight into the swings of fashion and consensus. What I outline here is the model that now has wide support; it provides a framework within which to plan further observations and to judge the significance of each piece of evidence. I also briefly review other models to show why they are not currently in favor, or at least why they are not thought to account for major sources of CR.

Supernovas are now generally considered to be the best candidates for CR sources, either directly or indirectly. The evidence is circumstantial but persuasive, comprehensive but not yet compelling. Some events that we now call supernovas were noted long ago, before there was any understanding of their true nature. The apparent permanence of the stars was deeply ingrained in ancient cosmologies. Occasionally, however, a star would suddenly appear where no star had previously been seen. This new star (called *nova stella* in medieval times) would brighten rapidly, then gradually fade. It was not until 1934 that the name *supernova* was used, by the Mount Wilson astronomers Walter Baade and Fritz Zwicky, to describe novas whose exceptional brilliance set them apart from all others. Baade and Zwicky suggested that supernovas were the sources of cosmic rays, although they thought that CR were mostly positrons coming primarily from supernovas in other galaxies.

The initial classifications of historically recorded events as supernovas rested on brightness alone, but other characteristics show them to be very different in their nature from the less luminous novas. Both novas and supernovas have been seen in our own galaxy as well as in many external galaxies. The relatively frequent novas have sometimes been considered as CR sources but are not currently favored.

No supernova has been seen in our galaxy since 1604, when Johannes Kepler and many other astronomers in western Europe witnessed the spectacular outburst of a new star in the constellation Serpens, not far from the positions of Mars and Jupiter at that time. At a place where no star was visible on October 8, a bright new star was first seen two nights later. It continued brightening until early November and then faded slowly, but it was still visible a year later. Records of extensive observations of the star's brightness, made by astronomers in Europe, China, and Korea, are in good agreement. At its maximum, SN 1604 was as bright as Venus. Among the other early supernovas are those seen in 1006, 1054, and 1572, the last one described in great detail by Tycho Brahe, the Danish astronomer.

All these early supernovas preceded Galileo's invention of the astronomical telescope in 1609. Astronomers did not have the opportunity to examine a really close supernova with modern telescopes and spectroscopes until the outburst in February 1987 in the Large Magellanic Cloud, a relatively nearby galaxy. Unfortunately, this object is in a part of the sky where it cannot be seen by the major northern-hemisphere observatories, optical or radio. Nevertheless, it is being followed over the entire e-m spectrum, and a special expedition went to Australia to carry out gamma ray observations with balloons. The measurements made so far give strong support to the standard theory of nucleosynthesis in supernovas. The steady decline of the visual brightness of SN 1987A suggests that the source of this light is a radioactive isotope of cobalt, ^{56}Co, produced by the decay of the nickel isotope, ^{56}Ni, that was synthesized in the explosion. Satellite observations have detected the expected high-energy X-ray photons.

Astronomers are on the alert for new supernova explosions in our galaxy, but in the meantime we have to rely on records of past events. The historical records from European countries tend to be spotty, but those in China, Japan, and Korea are often extensive. China had court astronomers almost continuously for 2,000 years, until 1911, and they made and recorded copious observations of the heavens. Transient events were noted for astrological purposes; appearances of what the Chinese called "guest stars" are of particular interest, now recognized as including meteors, comets (with and without tails), novas, and supernovas. The completeness of their records is exemplified by the inclusion of every return but one of Halley's comet since 240 B.C.; in addition, the supernovas of 1006, 1054, 1572, and 1604 are all well documented.

In contrast, there are no records whatsoever of any European observations of the 1006 and 1054 supernovas. These events, which were far too bright to have been missed, must have been seen but may not have been recorded because of the strong theological bias against acknowledging that any changes took place among the stars. Halley's comet and other comets were recorded, but they were considered to be atmospheric phenomena whose occurrence posed no threat to the prevailing dogma. It was only in the late sixteenth century that Tycho Brahe proved that a comet was not in the atmosphere but in fact was much farther away than the moon.

Eastern records have been scoured by modern astronomers, eager to identify other supernovas. Optical, radio, and X-ray telescopes have all scanned those parts of the sky in which putative supernovas were noted, and the astronomy of supernova remnants (SNR) has developed into an active field. In all of the clearly identified SNR, radio astronomers have detected synchrotron radiation emitted by energetic electrons as they spiral in a magnetic field. The electrons are usually confined in a roughly spherical shell, containing material expelled at high speed from the site of the original stellar explosion. This clear indication of high-speed electrons is one reason for considering supernovas as possible CR sources.

Until the appearance of SN 1987A, by far the best studied of all of the SNR was the Crab nebula, located at a distance of 2 kiloparsec (6500 light years) from Earth. For many years the Crab was little more than a curiosity. Charles Messier listed it as the first object in his 1771 catalog, designed to identify hazy objects that astronomers should ignore in their search for the comets that were then fashionable. (This was soon after the first return of the comet that Halley had so spectacularly predicted for 1758.) Interest in the Crab increased markedly after its identification in 1948 as the first galactic emitter of both radio and visible radiation. In the first radio survey of the sky, the Crab was among the brighest objects listed.

From analyses of the shape of the Crab's spectrum and its polarization, it became clear by 1955 that much of the Crab's radiation at visible wavelengths was synchrotron radiation, produced by electrons with energies above 10^{11} eV. The synchrotron character had been predicted by the Russian theorist Iosef Shklovsky, but he did not have access to a large enough telescope to make the observations needed to test this. It took several years before Walter Baade, using the new 200-inch telescope on Mount Palomar, discovered the Crab's polarization. Figure 8.1, taken a few years later with a Polaroid filter that was rotated between successive photographs, shows this effect, a characteristic feature of synchrotron radiation.

The discovery of the synchrotron nature of the Crab's visible

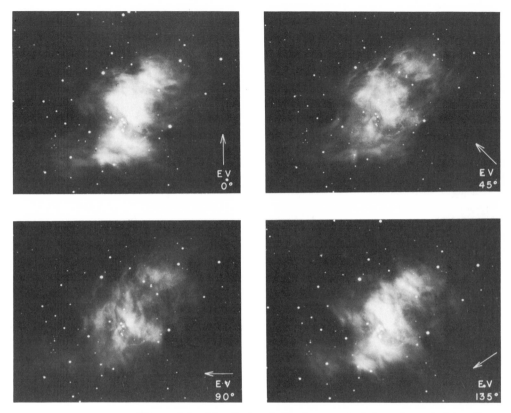

Figure 8.1   *The Crab nebula, photographed through a polarizing filter set at different angles. The changes seen in this set of photographs show that much of the light is synchrotron radiation, produced by high-energy electrons spiraling in a magnetic field. (Photograph courtesy of Palomar Observatory.)*

light introduced a problem that had no ready solution. Electrons with the needed energies would lose that energy so rapidly that after less than 100 years they no longer could radiate at visible wavelengths. Yet it was obvious that even after 900 years, the Crab was still a potent synchrotron source. A continuous supply of high-energy electrons was required, but none could be identified.

By now the Crab has been observed at all wavelengths, from radio to gamma rays. The synchrotron radiation emerges from a glowing, irregularly shaped region, about one-tenth of a degree across, corresponding to 10^{19} cm, more than half a million times the Earth-sun distance. The abundance of many elements has been spectroscopically analyzed and the debris identified as containing products of explosive nucleosynthesis. At the center of the Crab is a pulsar. This remarkable object, identified only in 1968, is rotating thirty times each second, beaming out its signals (in all wave-lengths) like a lighthouse. We think the rotational kinetic energy of the pulsar replenishes the electrons, but the mechanism remains elusive.

After it exploded in 1054, the Crab was bright enough to be seen by day for 23 days and by night for nearly two years. At the maximum, its luminosity was about 10^{41} ergs/sec, about one hundred million times that of the sun. Today, more than 900 years later, the energy being radiated by the Crab, from several different mechanisms and through all its different wavelengths, still amounts to 2×10^{38} ergs/sec—almost 100,000 times more than the sun radiates. The total energy radiated over the 900 years amounts to 10^{50} ergs. Today the remnants of the Crab are moving away from the explosion center with speeds of around 1,500 km/sec; the kinetic energy of the outward-bound debris amounts to around 10^{49} ergs.

If cosmic rays are produced mainly by supernovas, perhaps like the Crab, how frequently do such explosions occur? This is not easy to estimate. Our view of large regions of the galaxy is obscured by interstellar dust, and we cannot even see clearly to the galactic center. Radio waves help to map SNR once we know they are there, but discovery of new supernovas rests primarily on visual sightings, and the most recent event in our galaxy was the 1604 supernova observed by Kepler, Galileo, and many Chinese astronomers. From historical records and modern observations of other galaxies where supernovas have been seen, it is now estimated that one supernova explosion occurs every fifty years in our galaxy, but this rate could be off by as much as a factor of two.

The picture is further complicated by the fact that there are two main categories of supernovas. Type I supernovas are somewhat more abundant and have greater luminosities than those of Type II, and their brightness diminishes more rapidly. Type II supernovas probably represent the final stages in the evolution of stars appreciably more massive than the sun. SN 1987A is classed as Type II and may have involved a star initially of about twenty solar masses. Less than half of such a star might remain after the explosion, in the form of a rapidly rotating neutron star (Figure 8.2). The possible existence of this strange type of object, first suggested on theoretical grounds by Lev Landau in Russia in the early 1930s and by J. Robert Oppenheimer in the United States a few years later, is about 10 km across and has a magnetic field about 10^{12} times stronger than that of the Earth. The combination of rapid rotation and strong magnetic fields provides the ingredients for accelerating particles to CR energies. When pulsars were discovered, in 1967, there was at first no idea of their true nature. Oppenheimer's cal-

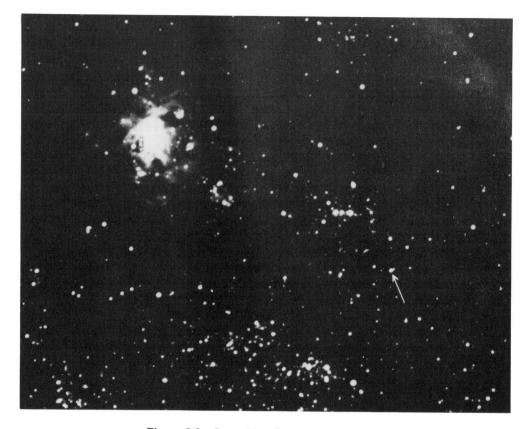

Figure 8.2   *Large Magellanic Cloud (LMC), shown before and after the outburst of Supernova 1987A. (left) November 19, 1960; (right) February 24, 1987, within a day after the explosion had been discovered. The brightness of this supernova was so great that a shorter exposure was needed, as can be seen by the other images in the second view. (Photographs courtesy of South African Astronomical Observatory.)*

culations had dealt with the nature of the collapsed star and not primarily with their radiation. The idea of identifying pulsars with rotating neutron stars was put forward by Thomas Gold of Cornell in 1968 and is now generally accepted.

The Type I supernovas are not as well understood. They are thought to occur in the late stages of evolution of a binary system, where two stars move in close orbits around each other; one of the stars is a white dwarf and its partner a normal star with an extended atmosphere. Material is attracted gravitationally to the surface of the dwarf star, where it accumulates. There is a well-established theoretical limit for the mass of a white dwarf: the *Chandrasekhar limit*, at close to 1.4 solar masses. Accretion of matter onto the dwarf ultimately produces a thermonuclear runaway when this

limit is exceeded. No central remnant has been found in Type I SNR, and identification rests on radio and visible detection of the expanding shell of debris.

The supernova model for the origin of cosmic rays becomes more attractive when we examine the energy budget. Within the solar system—the only region directly accessible to us—the measured flow of cosmic rays shows that, on average, each cubic centimeter has an energy content of 1 eV, and we presume that this is typical of the rest of the galaxy. With a galactic disk volume of 10^{67} cm^3, the total CR energy content is then about 10^{67} eV or 1.6×10^{55} ergs. As we saw, the ^{10}Be isotope abundance indicates an average CR lifetime of ten million years (3×10^{14} sec). Thus the rate at which CR energy is lost amounts to $1.6 \times 10^{55}/3 \times 10^{14}$ or 5×10^{40} ergs/sec. The energy requirements change if some of the galactic halo volume is included, but this is still a good first estimate. If the galactic CR energy is to be maintained at a steady value, this loss must be compensated by injection and acceleration of new cosmic rays.

If we accept the rate of one supernova in fifty years (1.5×10^9

sec) and assume that the typical supernova yields 10^{50} ergs in fast particles, then the average power, the rate of providing energy, is $10^{50}/1.5 \times 10^9$ or 6×10^{40} ergs/sec. Although this calculation is admittedly very rough, it does provide a plausible agreement in terms of the availability of energy from supernovas: these enormous amounts of energy are easily adequate to power cosmic rays.

Identifying supernovas as likely CR sources and demonstrating the adequacy of the energy they provide still leaves a major puzzle to be solved: how, in detail, is such a large fraction of this energy directed into CR? The enormous amount of energy released in supernovas comes mainly from gravity, during the process of contraction from a star with a typical radius of 10^{11} cm or more into a tiny remnant neutron star core having a radius of around 10^6 cm (about 6 miles).

The mechanisms responsible for CR acceleration are still unclear. In the Crab, diversion of energy to CR takes place with an efficiency of more than 1 percent. In contrast, the highest efficiency achieved with accelerators such as those at Fermilab (near Chicago) and CERN (near Geneva) is a hundred times less. We detect e-m signals at many wavelengths telling us that fast particles are streaming around the galaxy. What we do not yet know is whether any one type of source dominates CR production and acceleration, or, what is more likely, whether contributions come from several different sources and mechanisms.

In general, as seen in the laboratory, combinations of high voltages and strong magnetic fields can accelerate charged particles. Such acceleration occurs in several identified regions, for example in the Crab and around pulsars, in sunspots, and in the solar wind. Supernovas represent one source type in which rapid acceleration occurs in a relatively compact region. The ingredients are there, one might say, but the recipe evades us. On a larger scale, CR can be accelerated by moving magnetic fields. In this process, first suggested in 1949 by Enrico Fermi, the random motion of CR relative to a moving magnetic field results in the particles' sometimes gaining energy and sometimes losing it. On balance the gains exceed the losses, and a statistical gain occurs with a slow overall acceleration. With the Fermi mechanism, energy gains are proportional to a particle's momentary energy, but radiation losses increase with the square of that energy. This sets a limit to the maximum energy that can be attained, which is not a serious problem for protons and heavy particles but is a problem in the case of electrons, which radiate strongly because of their small mass. High-energy electrons can be produced through meson decay because the mesons are created with high energies, the electrons do not themselves require acceleration.

Supernovas can also contribute to acceleration over an extended region, as shock waves spread out from their explosions into the interstellar medium. Shock acceleration of particles has been observed, but on a much smaller scale, in solar flares and in the interaction of the solar wind with the magnetic fields surrounding the Earth, Jupiter, and Saturn.

Energy spectra near the Earth have been shaped by a combination of injection, acceleration, and propagation; these processes are modified by energy losses, collisions, and perhaps leakage from the galaxy, and they all depend on particle energies. Observations of solar CR from space probes and theoretical studies of acceleration mechanisms show that different acceleration processes can all produce CR with energy spectra having the characteristic E^{-a} shape. Many different models have been examined, but none is clearly the best candidate.

The combination of the chemical composition and the relative constancy of the energy spectra up to about 10^{14} eV suggests either that CR production is dominated by one type of source located within our galaxy, or that a blend of processes has smoothed away different histories. For particles above that energy, and certainly above about 10^{18} eV, the evidence favors assigning an increasing fraction of the cosmic rays to sources in external galaxies. Intergalactic CR exchange is a natural consequence of the inability of the weak galactic magnetic field to contain ultra-high-energy particles.

Another puzzle in the search for acceleration mechanisms is how the highest-energy particles, those above 10^{18} eV, are produced. I noted in Chapter 6 the evidence for a change in the shape of the energy spectrum at that high energy, and this can be linked to another observation not previously introduced. Once we allow for the effects of the Earth's magnetic field, most cosmic rays are found to arrive uniformly from all directions. For particles with high energies, this feature is studied by observing the arrival directions of air shower primaries. At energies up to 10^{15} eV the variation in the number of particles arriving from different directions is less than one-tenth of 1 percent. Examination of the arrival of showers with energies above 10^{17} eV has shown that isotropy is no longer the case: there is an excess of about 1 percent of showers whose axes point in the direction of a cluster of galaxies in the constellation Virgo. This evidence must be used with caution, however, since most air shower data come from the northern hemisphere, and the smaller pool of data from the southern sky is inconclusive.

At one time, as exemplified by the proposal of Baade and Zwicky, it was suggested that most cosmic rays came from outside the galaxy, but this idea was abandoned when the CR energy density was better evaluated and distances to other galaxies were more reliably determined. Potential CR sources are adequate to provide the energy to fill our galaxy with cosmic rays, but the energy problem would become unmanageable if the much greater volume between galaxies also had to be filled by CR to the same extent that we detect near Earth. An escape from this difficulty is provided if the only extragalactic CR are the very few with the highest energies.

There are additional reasons for no longer considering an extragalactic origin for most of the cosmic rays. First, the fragmentation of heavy particles and the resulting elemental and isotopic abundances would be different from those that we observe. Second, with the huge distances that would have to be traversed, the energy of the electrons would be degraded by collisions with many low-energy photons left over from the big bang. Most of these photons have wavelengths in the radio and infrared regions of the spectrum, and in fact the 1965 discovery of this radiation provided the single most important pillar for the big bang theory at that time. In the early stages of the big bang, the temperature was about 10^{12} degrees. Now, about 15 billion years later, the universe has expanded and the radiation cooled to barely three degrees above absolute zero. This is also cosmic radiation (though it differs from what we have been considering), and its photons are everywhere, in and out of galaxies. Within galaxies the distances are small enough that photon collisions do not usually concern us, but they cannot be ignored when we start to consider CR traveling over intergalactic distances.

A third argument against extragalactic origin of most CR is provided by high-energy gamma rays, described in more detail in Chapter 9. A major source of gamma-ray photons above 70 MeV is the decay of neutral pions, produced in the collisions of protons above 700 MeV with the ambient interstellar hydrogen. Satellite mapping has shown a strong concentration of these gamma rays toward the galactic equator, whereas a more widely spread distribution would be expected if the primary protons were largely extragalactic.

Our search for an origin of the cosmic rays started with the set of abundances and the energy spectrum, from which we deduced the distance traveled, the age of the radiation, and the injection energy requirements. Although supernovas in our galaxy appear to serve as a good first-order model, there are still several loose ends. The abundances present a problem when they are examined in greater detail. The steady improvement in particle identification and statistics, as achieved most recently by the satellite experiments on HEAO-3, has shown that the charge spectrum for particles with charges 26 and above is not consistent with what would be expected on the basis of r-process nucleosynthesis alone. Instead there is a generally good agreement with the relative abundances found in the solar system, which represent a mixture of *r*-process and *s*-process contributions. The solar system almost certainly formed when a dense region of the interstellar medium (ISM) contracted, and we do not know of any compact sources that are injecting a spectrum of solar-system-type particles. It seems, therefore, that we must search for an acceleration mechanism that works on general interstellar material.

In a qualitative sense, we expect acceleration to occur where supernova shock waves spread out and encounter atoms in the ISM, but quantitatively this model has a long way to go. A potentially supporting piece of information was recognized in about 1980. The interstellar material consists mostly of atoms, with some molecules and with only a modest degree of ionization, but cosmic rays are fully ionized nuclei without any attached electrons. It is widely assumed that the electrons are stripped from atoms during acceleration, for neutral atoms cannot be accelerated to CR energies. The energy needed to remove an electron from an atom is a few electron volts; the amount depends on the type of atom and how many electrons have already been removed. The energy required to remove the first electron is known as the first ionization potential (FIP), and there is a somewhat better correlation between CR abundances and FIP than with general solar system abundances; that is, where the FIP is high, there are fewer cosmic rays, and vice versa. This seems to indicate that atoms with lower FIP are preferentially accelerated because it is easier to ionize them and thus to provide the electric charge handle by which they can be linked to the magnetic field in an expanding shock. At present this observation remains somewhat enigmatic.

Another anomaly (in a picture with many anomalies) is provided by some of the data on isotope abundances. Cleaner identification of isotopes of neon (Ne), magnesium (Mg), and silicon (Si), provided by experiments on the ISEE-3 spacecraft, has shown excess abundances (at 60–80 percent levels) for the isotopes ^{25}Mg, ^{26}Mg, ^{29}Si, and ^{30}Si when compared to regular solar system abundances of the dominant ^{24}Mg and ^{28}Si. Stan Woosley at the University of California at Santa Cruz and T. A. Weaver at Lawrence Livermore Laboratory have pointed out that these data could be reconciled if the CR sources resided in stars whose initial C, N, and O abundances were 80 percent greater than the solar system material; in that case the subsequent nucleosynthesis would produce the observed abundances. This interesting (but ad hoc) assumption unfortunately cannot yet explain the observation of the ^{22}Ne isotope among cosmic rays, where it is four times as abundant as in solar system material. What is shown by this sort of observation and calculation is that, with ingenuity, many of the initially discrepant numbers can be "explained," but one is often left with significant puzzles. Improvements in nucleosynthetic theory may yet resolve the isotope problem, or it may be that the accepted baseline of solar system abundances is not the best guide for clues to sources.

Are other CR sources possible? There are a hundred billion stars in the galaxy, of many different types, and several of these types are possible candidates. Serious difficulties often quickly emerge, however. The large number of solar-type stars, for example, will surely have solar-type flares, but even during their most intense phases these cannot provide enough particles with high enough energies to account for CR. Rare stars, such as the Wolf-Rayet type, seem attractive as possible sources because they are extremely luminous and are losing significant amounts of mass, but we do not know whether they can or do produce CR. Ordinary novas probably produce some particles, but the numbers and energetics argue against their playing a major role. Another possibility is the galactic center, the scene of enormous activity as revealed by infrared, X-ray, and gamma-ray telescopes, but this is too far from the solar system to contribute much to CR observed near Earth. And, of course, there are black holes, theoretically predicted as final stages in the evolution of massive stars but not yet convincingly detected. These should be capable of providing sufficient energy, but the connection with cosmic rays has yet to be made.

Thus, by exhaustion of alternatives, we return to the supernovas. Where does this leave us, and what can we safely consider as well established? From the charge spectrum and the necessary fragmentation, the average 5 g/cm^2 path seems well established. From the ^{10}Be isotope, the lifetime also appears secure. The energy requirement rests on the reasonable assumption of galactic confinement (for most cosmic rays). Supernovas seem to be involved, although we are not sure exactly how. Some interstellar material is probably accelerated to mingle with those particles produced in compact regions. The highest-energy particles (probably protons) come from outside the galaxy, but the acceleration process has not been identified.

Although there are still numerous remaining loose ends, we have made progress. From the observed charge spectrum, we have been able to calculate the average distance traveled by the cosmic rays. The relative abundance of the ^{10}Be isotope has been used as a clock, and in this way, with the observed energy spectrum, we have been able to determine an energy budget. The combination of these different pieces of information has led us to the conclusion that most of the cosmic rays must be moving within the galaxy, and that only some of the relatively few with the highest energies can be extragalactic. But we have not yet made use of the information on electrons, which, as we will see in the following chapter, supports the picture we have been developing.

Cosmic Rays with Little or No Mass

So far I have focused almost exclusively on nuclear cosmic rays—protons, helium nuclei, and heavier particles. Protons carry most of CR energy, but heavy particles give information on composition and propagation. Electrons, as we will now see, can be exploited to yield complementary information. We are initially confronted with the same questions that arose with nuclei: where do the particles originate, and what happens to them as they roam around the galaxy? The fate of the electrons hinges on major differences between their behavior and that of nuclei. Electrons do not combine to form heavier electrons, nor can they be fragmented. Because of their small mass, electrons are far more susceptible to losing energy in ways that produce radiation detectable at Earth. I will begin by cataloging the most important ways in which CR electrons can be produced and can lose energy.

Electrons are injected along with protons and heavier nuclei in outbursts such as supernova explosions. Shock wave acceleration of interstellar particles acts on electrons as well as on nuclei. Electrons picked up by these processes will all be negatively charged—the same type of electrons as in all normal atoms. To these are added electrons generated in interstellar collisions of nuclear particles during their propagation. The most plentiful of these collisions involve CR protons with interstellar hydrogen nuclei, usually referred to as p-p collisions. (The contribution from heavier CR primaries is similar to that of the protons, but rather smaller.) In these collisions, much of the kinetic energy of the projectile proton is converted into roughly equal numbers of positively and negatively charged pions. In about a fiftieth of a microsecond the pions decay into muons, which, after about two microseconds, decay into electrons and positrons. Meson production in this type of collision has been thoroughly studied at many accelerators, re-

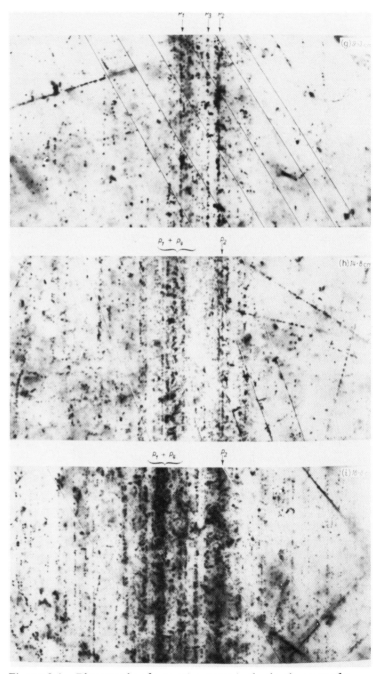

Figure 9.1   *Photographs of successive stages in the development of a shower with energy of 10^{15} eV. Photons produce electrons, which in turn produce more photons until the energy of each particle is too low for further multiplication. (Photograph courtesy of Peter Fowler.)*

vealing the energy spectrum of mesons and how their numbers increase with collision energy.

Cosmic ray "electrons" detected near the Earth include both primary and secondary electrons (e⁻) and secondary positrons (e⁺). In detecting "electrons," we do not always distinguish between the electrons and the positrons. Identification is often made by their common property, the production of an "electron"-photon cascade (Figure 9.1). The sign of the charge can be determined only in devices that use magnetic fields. It is now known that 10 to 20 percent of the primary "electrons" are positrons; the proportion depends on the energy.

As with the nuclei, the energy spectrum of the electrons (shown in Figure 9.2) can be represented by E^{-a}, and the changes of a with energy are clues to the history of the electrons. Electrons are far more sensitive than nuclei to the modulating effect of the solar wind. At energies below about 10 GeV, few traces are left near the Earth of the true interstellar electron spectrum, and it has been estimated (within large uncertainties) that the interstellar intensity of 100-MeV electrons is perhaps a hundred times higher than that seen near the Earth. At energies from 10 GeV and up, the modulating effect is not more than 10 percent, and interstellar inferences can be more safely drawn. At 10 GeV the energy spectrum has a slope of about 3, steepening to a value of 3.3 at higher energies. The overall electron intensity near Earth is a few percent of the proton intensity.

It is how electrons lose their kinetic energy that differentiates them sharply from the nuclei, and the basic factor is their small mass. Given equal forces, an electron will accelerate nearly 2,000 times as much as a proton. The acceleration or deceleration of a charged particle is always accompanied by radiation emission. For example, the sudden deceleration of electrons striking a metal target produces the X-rays used in medical radiography. For an electron moving along a curved path in a magnetic field, the radiated energy comes from the particle's kinetic energy, and synchrotron radiation is emitted (Figure 9.3).

For electron energies around 100 MeV, in a typical galactic magnetic field of 3 microgauss, synchrotron radiation is the dominant means of losing energy. The radiation emitted by each electron is spread over a range of frequencies but tends to peak at a frequency that depends on the strength of the magnetic field and on the square of the energy. Under these conditions, electrons of 100 MeV will radiate strongly at 0.48 MHz, those of 1 GeV at 48 MHz. Radiation at the lower frequency is difficult to detect because of the Earth's ionosphere, but 48 MHz is well within the range of radio astronomy. When synchrotron radiation is produced by many electrons whose energy spectrum follows the familiar E^{-a} form,

Figure 9.2
Electron energy spec-
trum. At energies be-
low about 10^9 eV, the
intensity and shape of
the spectrum are
strongly influenced by
the sun.

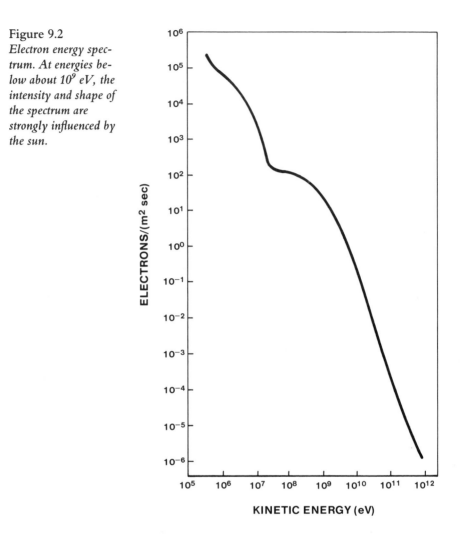

Figure 9.3
Electrons travel in
spiral paths around
magnetic field lines,
continuously radiat-
ing photons whose
energies depend on
the electron energy
and the magnetic field
strength.

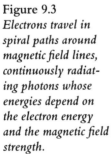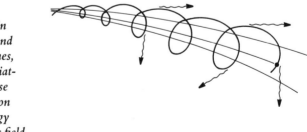

then the resulting radio spectrum will have a characteristic shape from which the value of *a* can be deduced.

Cosmic radio waves were first detected accidentally by Karl Jansky in 1931, while he was studying radio static generated in thunderstorms (Figure 9.4). Jansky found that his newly discovered "signals" were not produced on Earth but instead seemed to be coming from the direction of the center of the galaxy. Radio astronomy mushroomed after World War II with the major improvements in electronics, and it quickly became apparent that the cosmic radio "noise" (as it came to be known) had a totally different spectrum from the thermal spectra expected from astronomical objects. This "nonthermal spectrum" was later identified with synchrotron radiation.

By using synchrotron formulas and mapping the cosmic radio spectrum, we can, in principle, deduce the energy spectrum of electrons whose radiation we are detecting. The actual analysis is somewhat complex. Electron acceleration can also be caused by an electric field such as an electron experiences briefly when it passes close to an atomic nucleus. The resulting radiation is known as *brems* (braking) -*strahlung* (radiation). Bremsstrahlung from a multi-MeV electron can produce X-ray and gamma-ray photons. In interstellar space, CR electrons will typically travel a distance about the size of the solar system before having a sufficiently close encounter with an atom to produce bremsstrahlung. The result can be catastrophic for the electron, which can lose almost all of its energy in a single collision. In contrast, synchrotron radiation extends over galactic dimensions, and the electron radiates continuously. Each of the emitted synchrotron photons has low energy, and the electron's kinetic energy slowly trickles away. Bremsstrahlung is the most likely origin of low-level background gamma-ray photons, as has been mapped by two satellites, NASA's SAS-2 and the European COS-B.

A third way in which electrons can lose energy is known as *inverse Compton scattering;* and it is most significant for the highest-energy electrons, above about 1,000 GeV (10^{12} eV). In this process, a high-energy electron collides with (is scattered off) a photon—the inverse of the effect that Arthur Compton discovered in the early 1920s. Inverse Compton scattering has turned out to be important in several astrophysical settings. The many photons in interstellar space have come from two major sources. On average, there is about one starlight photon with an energy of a few electron volts in every 10 cm^3. Their energy density amounts to about 0.6 eV/cm^3, quite close to the energy densities of cosmic rays and the galactic magnetic field. Many more photons are relics of the big bang cosmological explosion. Their energy density is about 0.3 eV/cm^3, but since they have low average energy (about 0.0013 eV),

Figure 9.4   *(left) Karl Jansky and the radio telescope that he designed in 1929. This antenna was 100 feet in diameter and was tuned to a wavelength of 14.6 m (20.5 MHz frequency). Located at Bell Telephone Laboratories in Holmdel, New Jersey, the whole system was mounted on four Ford Model T wheels and rotated once every 20 minutes. (right) Modern radio antenna: twenty-seven of the dishes can be moved on rails that extend along the three arms of the Very Large Array (VLA), located near Socorro, New Mexico. Each of the VLA arms is about 20 km long, and each dish has a diameter of 25 m. The spacing of the dishes can be changed, depending on the measurements that are to be made. (Photographs courtesy of Bell Telephone Laboratories and NRAO/AUI.)*

there are around 200 of them in every cubic centimeter. These photons are found everywhere—all over the galaxy and also between galaxies.

Inverse Compton collisions between electrons and cosmological photons can drastically reduce the electron's energy; the recoiling photon absorbs energy to appear as a gamma-ray photon. This type of collision effectively limits the observable electron energy spectrum to be around 3,000 GeV. At higher energies, electrons are degraded so rapidly by inverse Compton collisions that few will ever be seen at Earth. Indeed, the highest-energy electrons observed are in the region of 1,000–2,000 GeV.

To construct a general model for the CR electrons, we rely on direct observations of electrons and positrons near the Earth and add our detailed knowledge of the production of mesons in p-p collisions. From radio astronomical surveys, we know the spectrum of the CR noise emitted by distant electrons whose energy spectrum we can also calculate. We can allow for other ways in which electrons lose energy and adapt the leaky-box confinement model created for nuclear particles. What then emerges is a plausible picture. At injection, electrons have an energy spectrum proportional to $E^{-2.3}$. Because energy losses vary with electron energy, the emerging synchrotron spectrum will not have a simple form, but there is satisfactory agreement with what the radio astronomers observe.

Primary protons with energies above 10^{15} eV can produce mesons and therefore electrons with energies well above 1,000 GeV, but there will be few of these because so few protons have the necessary energy. Even with such high electron energies at production, inverse Compton losses produce an effective cutoff to the electron spectrum. Synchrotron losses dominate down to energies of 10 GeV; at lower energies escape from the galaxy becomes important. Below about 1 GeV, the solar wind modulation is so large that we lack reliable information on the interstellar electron intensity, which is perhaps 100 times larger than what we see near Earth.

A good fit between the synchrotron radiation spectrum and the near-Earth electron observations is obtained by using an average interstellar magnetic field of 5 microgauss and a lifetime that varies between 30 million years at 1 GeV and 4.8 million years at 100 GeV. These are rather longer times than are generally deduced from the ^{10}Be data. The observed electron/positron ratio is consistent with production in p-p collisions, provided the proton path length is about 10 g/cm^2, which is somewhat larger than that deduced from production of L nuclei by C, N, and O nuclei.

Further observations and modeling are clearly needed to reconcile the differences between quantities such as field strength or path length as deduced from different observations. The injection processes still remain a large area of uncertainty. We know that electrons are accelerated in the Crab and other supernova remnants, with and without pulsars, and, on a much larger and more distant scale, we detect synchrotron radiation emerging from quasars (probably cores of active galaxies). But the injection process continues to elude us.

Unlike everyday objects and atomic particles such as nuclei and electrons, photons do not have any mass in the conventional sense. They do not respond to Newton's formula $F = ma$, for no force can accelerate a photon; its speed is always that of light. Still, photons carry energy and momentum, and these can change when photons collide with particles.

The distinction between particles, photons, and waves has been blurred for many years. This issue has provoked controversy since at least the late seventeenth century, when Newton and Huygens championed competing particle and wave theories to explain the nature of light. Quantum mechanics and the less well known quantum electrodynamics constitute a comprehensive modern theoretical framework that has been extraordinarily successful in quantitative application, although we often (for convenience) use the older and simpler particle or wave models for specific problems.

Where, then, should we draw the line in defining cosmic rays? By now they are generally thought of as particles, although a photon model was strongly advocated by Millikan. If we were to include all photons in the definition, we would have to include all of astronomy, which is not practicable. Accordingly, we extend our boundaries to cover only the high-energy X-ray and gamma-ray photons produced by CR particles. In this way we can use photons as diagnostics of distant CR, as we already did for electrons with the radio spectrum.

Photons with much higher energies can also be traced to the creation of pions in interstellar p-p collisions. From the same collisions that produced charged pions, neutral pions also emerge and promptly decay into gamma rays detectable over a range of energies peaking at 70–100 MeV. While production of neutral pions in the Earth's atmosphere quickly leads to electron–photon cascades, in the extremely low density of interstellar space no such showers can develop. Photons from neutral pions therefore travel across the galaxy with little chance of absorption.

The most complete celestial surveys of high-energy photons have been carried out by two satellites, NASA's SAS-2 and the European COS-B, launched in the 1970s. SAS-2 detected 8,000 photons. COS-B, during its lifetime of nearly seven years, detected 210,000 photons, about as many as the number of less energetic photons we receive by eye in 40 seconds from the faintest visible star. The energy of each gamma ray photon was measured through the cascade it produced in a spark chamber. A celestial gamma ray map has been constructed (Figure 9.5), and it shows that the diffuse emission is clearly concentrated in the galactic disk. Several conspicuous compact regions of high intensity can be identified with known objects such as the Crab and the strong X-ray source Cygnus X-3. The intensity of diffuse high-energy gamma rays agrees with that expected if the general galactic cosmic intensity is similar to what we detect near the Earth.

Lower-energy gamma rays, in the range of a few MeV, can be produced by a different set of CR interactions. Protons with energies below 100 MeV cannot produce mesons, but when they collide with nuclei some of their kinetic energy is transferred and the nuclei become excited. When one of these nuclei returns to its normal state, the excess energy is emitted as a gamma ray with a sharply defined energy. (This differs from the situation with synchrotron radiation and bremsstrahlung, where photons can have any energy over a wide spectrum.) Proton collisions in dense regions around or between the stars are expected to yield gamma rays characteristic of certain nuclei and thus to be easily identifiable. In oxygen and carbon atoms that might be plentiful in dust grains, proton bombardment can produce gamma rays of 6.1 MeV and 4.4 MeV, respectively.

The most elusive CR particles are the neutrinos. Like photons, they have neither mass nor electric charge. They carry energy and momentum and can interact with electrons and nuclei, but seem to do so with the utmost reluctance. The likelihood of a particle's interacting is often expressed as a cross section, the effective target area it presents to other particles. For an atom, a typical cross

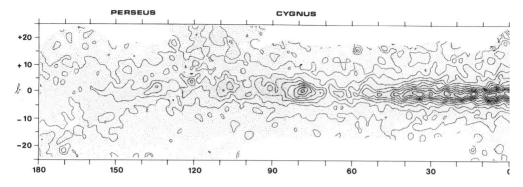

Figure 9.5   *The Milky Way Galaxy; large-scale distribution of galactic gamma radiation observed by COS-B satellite. The scale along the bottom of the figure indicates galactic longitude. Note the very strong source of gamma rays near 270°, identified as the Vela pulsar. The Crab nebula is at 185°. (From H. Mayer-Hasselwander et al.,* Astronomy & Astrophysics *105 [1982]: 164.)*

section is 10^{-15} cm^2; for a nucleus it is around 10^{-27} cm^2. The typical unit for capture of slow neutrons, leading to fission, is around 10^{-24} cm^2, which Fermi characterized as being as big as a barndoor—hence the term *millibarn* for 10^{-27} cm^2. For neutrinos of MeV energies, the cross sections run around 10^{-43} cm^2, but this is strongly energy-dependent. In general, a neutrino has only a few chances in a trillon of having a collision even while passing diametrically through the Earth. The process of detecting neutrinos is difficult and inefficient. Nevertheless, they are important members of the family of elementary particles, and the elucidation of their properties has kept a lot of high-energy physicists very busy.

Neutrinos are produced in several ways. Some are emitted in the decay of radioactive nuclei, while others come from mesons. When a pion decays to a muon, a neutrino is produced; later, when the muon decays to an electron, two neutrinos appear. Neutrinos are generated in large numbers during fusion reactions in the sun's core or in a supernova explosion. The galaxy is awash in neutrinos, which pass through us without any effect. Because of their relation to nuclear and particle processes, neutrinos should be considered as a component of cosmic ray physics, and only the technical difficulties of neutrino detection have slowed the development of this field.

Striking but puzzling results have come from two cosmic neutrino observations. For almost twenty years, Raymond Davis of the Brookhaven National Laboratory operated a solar neutrino detector in the Homestake Gold Mine in Lead, South Dakota, 1.5

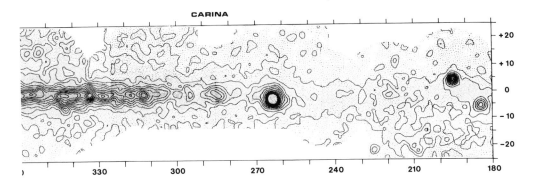

CARINA

km underground (Figure 9.6). Because of the expected low neutrino counting rate, this location was selected to minimize the CR
background. The detector's core was a tank holding 400,000 liters
(about 100,000 gallons) of extremely pure perchlorethylene, a
cleaning fluid. Solar neutrinos react with the fluid's chlorine atoms
to create ^{37}Ar, an isotope of argon, which is radioactive with a half-
life of 35 days. To detect the neutrino-produced argon, Davis
flushed the system with helium gas every 100 days to extract the
argon. During those 100 days many of the argon atoms decayed,
but the radioactivity of the survivors could be measured.

The sun's neutrino production can be calculated from the
theory of stellar structure, generally considered to be well established. Just over 10^{37} neutrinos leave the sun every second, and at
Earth the flux is about $10^{10}/cm^2$ sec. But for every 10^{21} neutrinos
that penetrate the neutrino detector, only one will induce a chlorine
atom to change to argon, and Davis's expected rate of production
was about three atoms per week. A *solar neutrino unit* (SNU) is
defined as the number of captured neutrinos per 10^{36} target atoms,
and in these units the predicted counting rate was 7.6 SNUs. But
Davis's observed rate, though it varied a bit, had an average over
the years of just under 2 SNUs, less than one-third of the predicted rate.

The importance of this discrepancy between observation and
theory cannot be overemphasized. Modern stellar theory has been
widely accepted, founded on principles that were thought to be
well tested. Because this theory has brought order to an otherwise
chaotic diversity of astronomical observations, there is great reluctance to abandon it. Yet there is clearly a problem when our closest
and best studied star refuses to conform.

All sorts of explanations have been considered. The accepted
solar model has been constructed to fit the observed luminosity and
size, both beyond question. But the light coming from the solar
surface originated more than a million years ago in the core's

Figure 9.6
Tank containing 100,000 gallons of the neutrino detector perchlorethylene in the Homestake Gold Mine, South Dakota. (Photograph courtesy of Raymond Davis, Jr., and Brookhaven National Laboratory.)

nuclear reactions. The neutrinos, on the other hand, because of their obliviousness to other particles, travel directly from the sun's core to Earth in 8 minutes. Could the neutrinos, in telling us what was occurring 8 minutes ago, be foretelling the surface behavior of the sun a million years in the future? Alternatively, could neutrinos be more complex particles than we presently believe? Several exotic possibilities have been considered and discarded for lack of supporting evidence.

Another puzzle relates to neutrino mass, long taken to be exactly zero, as originally assumed and since supported by experiments and calculations. If, instead, neutrinos had finite mass, there could be significant consequences, including one on a cosmological scale. The spectra of galaxies prove that our universe is expanding; the more distant a galaxy, the greater is its speed away from the Earth. What is not known is whether the universe is open or closed. Will it continue to expand forever, or will it, sometime in the remote future, stop and begin to contract with increasing speed, reversing the big bang to create a big crunch?

What will determine the fate of the universe is its total mass. With sufficient mass, gravitational attraction will ultimately slow down expansion and force a reversal into a contraction. There are so many neutrinos in the universe that their contribution to the critical mass can be significant if they each have some mass, even though very little. Numerous experiments have already shown that the neutrino mass is certainly less than 0.0001 times the electron mass, but that would still be large enough to close the universe. The challenge is to see whether this experimental limit can be pushed down closer to zero.

In 1987 when the new supernova (SN 1987A) appeared, scientists were able to detect neutrinos close to the time of the visual outburst. Neutrinos are reluctant to interact with other particles, and of the more than 10^{16} that streamed through detectors on Earth, only 19 neutrinos were detected, fewer than the number of scientific papers since written about them. In one neutrino detector system, operated near Kamioka, Japan, by a Japan–University of Pennsylvania collaboration, 11 neutrinos were detected over an interval of a few seconds. The other 8 neutrinos were detected in a system near Cleveland, run by scientists from the University of California at Irvine, the University of Michigan, and Brookhaven National Laboratory.

The precise time of the onset of optical flaring cannot be pinpointed to better than a few hours, so any delay of the neutrinos' arrival relative to the outburst is not well known. What can be pinpointed very precisely is the spread among the neutrinos arrival times. Particles with zero mass will all travel at exactly the speed of light. If the neutrinos were all emitted in less than a second (as suggested by some theorists), they should all have arrived within a similarly small interval. If, however, the neutrinos have nonzero masses, then, like electrons or protons, their speeds depend on their energies, and a spread in energies will show up as a spread in arrival times, as was seen. From the length of the interval over which the neutrinos were detected and from the measured neutrino energies, the neutrino mass can be calculated.

Different theorists, using different assumptions, have arrived at slightly different values for the neutrino mass, typically no more than 1/50,000 of an electron mass. This is a major discovery, for so small a neutrino mass is hardly sufficient to close the universe. The problem should not yet be considered as solved, however; it is possible that present assumptions about the supernova explosion itself, as well as the behavior of neutrinos within the exploding star, might change.

We are left with tantalizing appetizers, waiting for the next supernova explosion. It has often been remarked that astronomy is an observational but not an experimental science, and this is no better illustrated than by the neutrinos. The essential element of experimental science is missing: we cannot control the conditions of the "experiment." We are unable to adjust the sun's temperature and wait to see how this affects the flow of neutrinos; we cannot arrange for a supernova to explode in a star of known mass. Our ingenuity is confined to the design of our apparatus and the sophistication of the later analysis.

Chapter 10

The Subnuclear World

Cosmic ray studies have seen several shifts of emphasis. Currently most attention is directed toward the astrophysical aspects, but for a period of about two decades, cosmic rays provided the only way of discovering new subnuclear particles. Today, the study of these particles is carried out almost exclusively in a few large international centers such as the proton accelerators at the CERN laboratory in Geneva and at Fermilab near Chicago, and the electron accelerators at SLAC at Stanford and DESY in Hamburg. Occasional CR observations may indicate the presence of new particles in an energy range still beyond any machine's ability, but this is no longer where the action is.

When cosmic radiation was discovered, views of the structure of matter were simple. The catalog of the elements had been based on chemical separation of materials found in nature. Each atom of each element was supposed to consist of a central nucleus carrying a positive electric charge, surrounded by a number of negatively charged electrons. In this way, the atom as a whole was electrically neutral. Within the nucleus were thought to be protons carrying positive charges and a sufficient number of electrons to bind the protons together; otherise the collection of like-charged protons would soon disperse. No other (nonelectric) forces were known or invoked.

Although it did not provide the answers to all questions, this atomic model survived the invention of quantum mechanics in the 1920s and lasted up to James Chadwick's discovery of the neutron in 1932. Chadwick used alpha-particles emitted from radioactive polonium to bombard a thin foil of beryllium. Neutrons ejected from the beryllium nuclei were identified as they collided with protons whose recoils could be detected. Typical alpha-particles from radium or polonium have energies of a few MeV, however—

far below CR energies. Understandably, therefore, subsequent particle discoveries came first from cosmic rays and then from more energetic accelerators.

Working at almost the same time as Chadwick, Carl Anderson and Seth Neddermeyer of Caltech were encouraged by Millikan to pursue their study of the penetrating power of certain CR particles. In 1932 they used a cloud chamber surrounded by a giant electromagnet to identify CR particles and measure their energies; the discovery of positrons rested on seeing their curved paths in the magnetic field. It was just at this time that Blackett and Occhialini found, with a counter-controlled cloud chamber, roughly equal numbers of positrons and electrons and thus provided the first confirmation of Dirac's theory of antiparticles.

Thus by the end of 1932 the catalog of particles had grown. Nuclei were known to consist of protons and neutrons. Dirac's particle and antiparticle theory had passed a major test, and there was an increased theoretical attack on the problem of nuclear stability. The electric force between two particles depends inversely on the square of their separation. Within the tiny confines of a nucleus (less than 10^{-12} cm), this $1/r^2$ repulsive force between protons should increase to overwhelming proportions. What sort of force holds the charged protons together against their mutual repulsion and also binds them to the uncharged neutrons? Although there were several other major questions regarding the nucleus, the central one in the 1930s was still that of binding.

The key to the understanding of nuclear forces came from a new idea, that of *exchange forces*. An analogy used by the theorist Paul Matthews is more helpful than a definition: Imagine two children playing with a ball, tossing it back and forth. The distance they can stand apart will clearly depend on the weight of the ball. If they play with a tennis ball, they can stand far apart, but playing with a basketball keeps them much closer together. We can say that they are bound together by the exchange of their ball. At times A will hold the ball, at times B has it, and at times it is in transit, being exchanged. The *range* of an exchange force varies inversely with the mass of the exchanged object. In theoretical models of the electromagnetic force, the photon, that bundle of energy, is the exchanged object, the "ball" that holds an electron to its positively charged nucleus or causes the deflection of one charged particle passing near another. Because the photon's weight is zero, the range of the electromagnetic force is great. The force, with its $1/r^2$ dependence, extends to infinite distances, although its strength does diminish to a vanishingly small value.

Hideki Yukawa at Osaka University first considered the possible exchange of electrons between protons and neutrons but found that the resulting force was too weak to provide the needed nuclear binding. Early in 1935, Yukawa proposed instead an exchange particle whose mass of around 200 electron masses would ensure that the force would have a short range. With the resulting exchange force and with a more complex mathematical formulation than that for electromagnetism, Yukawa constructed a model for the behavior of nuclei. The new nuclear force was designed to be stronger than the well-known electromagnetic force at close distances but to decrease more rapidly as the distances increased because of the heavy particles being exchanged. At large distances the electromagnetic force would still dominate and account for the binding of atoms into molecules and molecules into solids and liquids.

Because the new particles appeared to have masses intermediate between those of the electron and the proton, they were first called *mesotrons*. The shorter term *meson* is now generally used, and in a more restricted way. Discovery of mesotrons came in 1937 when Anderson and Neddermeyer found CR tracks in their cloud chamber, which was operating in a strong magnetic field at their Pike's Peak laboratory located 4,300 m above sea level. The presence of mesons was shown by particles that had sufficient energy to pass cleanly through a 1-cm plate of platinum placed across the middle of their cloud chamber. The curvature of the track segment below the plate was markedly more tightly coiled than the segment above, indicating that the particle had lost energy in going through the plate. The curvature of the two track segments, along with the calculated energy loss in the metal plate, showed that the particle's mass had to be around 200 electron masses. An alternative explanation, that the energy loss theory was at fault at high energies, was not entertained for long because it raised more problems than it solved.

Anderson's discovery was informally announced in Stockholm in December 1936 when he was accepting the Nobel Prize, awarded for his discovery of the positron a few years earlier. Mesons were independently discovered by J. C. Street and E. C. Stevenson of Harvard University and by Yoshio Nishina, M. Takeuchi, and T. Ichimiya of the Institute of Physical and Chemical Research in Tokyo, during 1937. (It is interesting to note that Yukawa's theory, published in Japan, took several years to receive attention in the West. Anderson and Neddermeyer made no reference to Yukawa when reporting their discovery; even Nishina and his colleagues seemed unaware of Yukawa's calculations.)

Yukawa's particles had been devised to meet the specific needs of nuclear stability. Their masses were expected to be substantially greater than those of the electrons but still much less than those of protons or neutrons, and the cloud chamber tracks were hailed as confirming this bold hypothesis when Yukawa's theory became known. But that theory also required that mesons possess other measurable properties. A chance observation gave added confirmation, but then a dramatic failure posed a serious problem.

Although the mesons had been devised to act as a nuclear glue, they were expected to be radioactive (hence unstable) when not confined within nuclei. This sort of behavior might seem paradoxical, but it was already known to be the case for neutrons. Many nuclei that contain neutrons are perfectly stable—witness the world around us—but free neutrons live for only around 10 minutes. The discovery of meson decay was therefore not unexpected; it came from E. J. Williams and G. E. Roberts of the University College of Wales in Aberystwyth. These two researchers took a cloud chamber photograph showing the track of a fast electron starting from the end of the track of a very slow meson (Figure 10.1). Clearly, the meson had decayed, transferring its charge to the electron. Presumably one or more neutral particles must also have been produced in the decay but left no tracks.

The mesons that crossed the cloud chambers of Anderson and Neddermeyer, Williams and Roberts, and other researchers were generally moving faster than three-quarters of the speed of light, spending around one-billionth of a second in a chamber. The mesons' half-life had to be longer than this, but how much longer was not immediately clear, for the production sites of the mesons had not yet been identified.

The first experiments specifically designed to measure the lifetime of mesons were carried out by Bruno Rossi. By this time (1938) the political situation had forced him to leave Italy, and he held positions at the University of Chicago and Cornell University, where he made his measurements. Cosmic ray particles had been classified as either *soft* or *penetrating,* on the basis of their ability to pass through layers of absorbing material stacked above or inside the cloud chambers or between Geiger counters. The penetrating particles turned out to be mostly mesons, and Rossi measured their numbers at different altitudes, with and without absorber blocks above his counters. The underlying idea was that mesons, produced somehow by cosmic rays in collisions with atoms at all levels of the atmosphere, would move rapidly to lower altitudes. Along the way, their numbers would be reduced by decay and collisions. The number measured at any altitude would thus depend on a combination of those two removal mechanisms. Measurements at different altitudes, with and without extra absorber, would separate the two effects.

Rossi had to make some assumptions about the average kinetic energy of mesons. He also had to allow for an effect predicted by the special theory of relativity: because of their extremely high speeds, mesons would appear to live longer than their "true" lifetime (defined when measured at rest). Although this relativistic

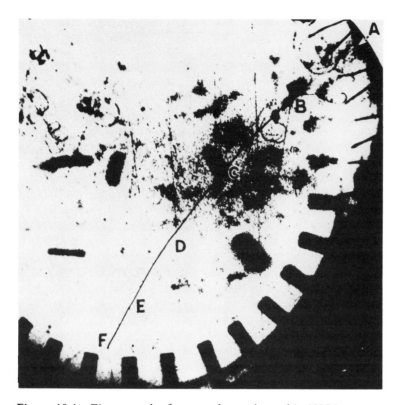

Figure 10.1 *First example of a muon decay, observed in 1937 in a cloud chamber by E. J. Williams and G. E. Roberts. The muon enters the chamber at A and travels to F, where it decays to produce a secondary electron, whose short and thinly ionizing track is barely visible. (Photograph courtesy of L. Thomas and D. F. Falla, University College of Wales, Aberystwyth.)*

effect has now been confirmed in countless experiments, Rossi's demonstration was the first and is still one of the most elegant. In the end, Rossi obtained a value for the meson mean life of about 2 microseconds.

In a second experiment, the lifetime was measured much more directly. Electrons were detected from mesons that had slowed down and come to rest in a block of absorber surrounded by many Geiger counters. The time interval between the meson arrival and the emergence of the decay electron was determined by a high-speed electronic circuit designed by Rossi; he obtained a mean lifetime of 2.15 microseconds—very close to the present best value.

By this time Yukawa's theory of meson properties had become known. It was expected that two versions of mesons would exist, something like the electrons and positrons with their identical masses but opposite electric charges. Traveling through the largely empty atmosphere, both positively and negatively charged mesons would decay, producing positive and negative electrons. In contrast, the behavior of mesons was expected to be quite different when they came to rest in dense concentrations of atoms. The positive particles should decay, but the negative ones should be attracted to positively charged nuclei and rapidly absorbed, long before they could emit their customary electrons. Absorption of a meson into a nucleus might cause that nucleus to split or to emit particles such as protons, neutrons, or alpha particles. A check on these predicted meson properties was the next order of business, and it was carried out by Marcello Conversi, Ettore Pancini, and Oreste Piccioni in Rome during the period 1943 to 1945.

Conditions in Rome in those days were chaotic. Conversi has described how he had been rejected for military service because of a vision problem, and thus was able to continue intermittently with his CR research. Piccioni was at one time arrested by the Germans while trying to link up with the advancing Allied forces, but was soon released. Pancini, at one time with the partisans, rejoined the group in 1945 after the liberation of northern Italy. Conversi, also involved with the growing Resistance movement, had moved the laboratory closer to Vatican City, where he thought it would be safer from Allied bombing.

Once things settled down, the three researchers were able to identify mesons according to their charges and look for their decay electrons. It became clear that the negative mesons were not being captured by nuclei in the way expected. Nuclear captures did occur when negatively charged mesons were stopped in iron absorbers, but not when the absorber consisted of the much lighter atoms of carbon. This result, described by the Rome group in their 1945 paper in the *Physical Review,* showed that the mesons so hopefully identified earlier with Yukawa's heavy exchange particles did not possess the critical quality of always interacting strongly with nuclei. They simply could not be the agents of the strong force to bind nuclei.

The discovery of positrons and mesons came through the use of cloud chambers, with the later addition of Geiger-Müller counters to provide the coincidences for triggering the chambers. The Rome experiment had used only counters. After 1945 improved Ilford photographic emulsions provided the experimental basis for the next discoveries, made in 1947.

To study the effects of cosmic rays, Powell and Occhialini at Bristol University used some of the new emulsions, which were left at high-altitude research stations such as those on the Pic du Midi in the Pyrenees, the Jungfraujoch in the Alps, and the Bolivian Andes. During exposures of some weeks, numerous CR tracks were accumulated, but the early emulsions suffered from fading—older tracks did not possess as many grains as more recently formed tracks, and quantitative data were initially uncertain. But a few tracks were conspicuously different from all the others. In each case a steady increase in the closeness of grains, and an accompanying increase in scattering, identified a track as having been produced by a particle that was coming to rest. From its end another track started, and it too displayed the characteristic signs of its particle slowing and stopping.

Grain measurements showed that each primary particle (labeled π, whence pi-meson or pion) had a mass of around 350 times the mass of an electron. The emerging secondary particle (originally labeled as μ, whence μ-meson or muon) had a mass in the range of 200–300 electron masses. The striking feature of these early tracks was that all of the secondaries (the μ's) had the same length, close to six-tenths of a millimeter. This indicated that all had started with the same initial energy, and the immediate conclusion was that only one neutral (unseen) particle could have been produced at the same time in each decay (Figure 10.2).

Unaware of the exciting new results being obtained in Bristol, Robert Marshak of the University of Rochester and Hans Bethe of Cornell proposed in the summer of 1947 a theoretical way of reconciling Yukawa's idea of the exchange force with the Rome observations of the mesons' reluctance to interact strongly. In the "two-meson" theory by Marshak and Bethe, the Yukawa meson was the agent of the strong nuclear force and was produced in violent nuclear collisions, but it lived for only a short time, and its longer-lived secondary was thought to be the CR meson already discovered by Anderson and later studied by the Rome group. Quite independently, Shoicho Sakata and Takesi Inoue had also developed a two-meson theory in 1943, but wartime conditions prevented these calculations from becoming generally known. The English translation of their paper reached the United States only at the end of 1947.

After the Rome experiments and shortly before the discovery of π-μ decays, the capture of mesons by nuclei had been directly observed. Both Don Perkins at Imperial College in London and the Bristol group found events in their emulsions in which the characteristic track of a slowing and stopping meson ended with several short black tracks. These little "stars" were produced by the capture of mesons by nuclei in the photographic emulsions. The nature of the captured mesons was not clarified until decay events were seen later that same year. It then became clear that the negatively charged pions were indeed being captured by nuclei, in contrast to the negatively charged muons, which often decayed.

Another discovery was made in 1947 by George Rochester and Clifford Butler of Manchester University while the two researchers were examining CR tracks in their cloud chamber. Most

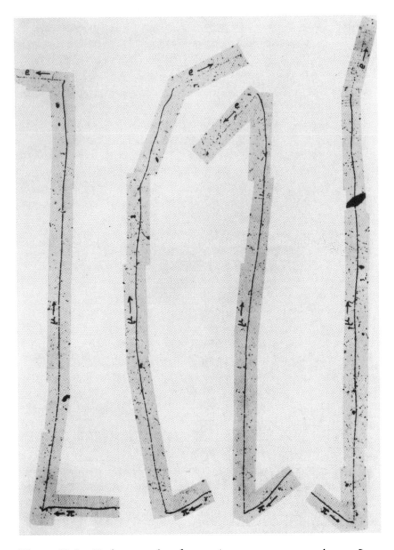

Figure 10.2 *Early examples of successive* $\pi - \mu - e$ *meson decays. In each of these events, only a short section of the pion track (π) is shown, leading to the place where it comes to rest. At that point the muon (μ) emerges with a kinetic energy of 4 MeV. After traveling for only 0.6 mm, the muon also comes to rest, and an electron (e) emerges to produce the characteristic low-grain-density track. (Photograph courtesy of Peter Fowler.)*

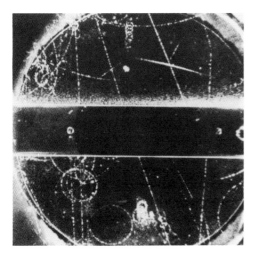 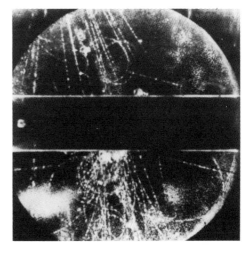

Figure 10.3 *First cloud chamber photographs showing tracks of V-particles. (left) The charged V-particle decayed at the top right, and the charged secondary was energetic enough to go clear through the lead plate across the chamber. The mass of the entering V-particle was at least 1,000 times the electron mass. (right) The neutral V-particle decayed just below the lead plate; the tracks of the two charged secondaries form the inverted V. The mass of the unseen neutral V-particle was estimated to be around 770 times the electron mass. In both photographs, other CR tracks stream across the chamber. Slow electrons can easily be identified by their strongly curved tracks. (Photographs courtesy of George Rochester.)*

of the tracks they observed came from particles that traveled clear across the chamber, but some interacted in a lead plate across the chamber's center. Rochester and Butler found two strange tracks that did not correspond to any of the previously known types of particle behavior. Both of these photographs, reproduced in Figure 10.3, showed tracks with abrupt bends, so the tracks looked like large V's. One of these looked like an inverted V, with its two prongs pointing down; the other looked like a V on its side. Many CR tracks do show sharp changes in direction where the particle has bounced (scattered) off the nucleus of a gas atom. At the point of such a scattering there should be a track produced by the recoiling target nucleus, but none was seen by Rochester and Butler. Another distinguishing feature of these V-particles was the ionization along each track. Ionization in a cloud chamber shows up in the number of small liquid droplets along a track. Although drop counting was not as reliable as the corresponding grain counting in photographic emulsions, Rochester and Butler could still draw some firm conclusions that precluded any conventional explanation.

The measurements of these two researchers showed that the

Incorporation of a lambda-particle within a nucleus introduces some exotic nuclei not otherwise found in nature. For example, hydrogen has three well-known isotopes: the dominant ^{1}H, with a single proton as the nucleus; heavy hydrogen or deuterium, ^{2}H, with a nucleus consisting of a proton and a neutron; and the radioactive tritium, ^{3}H, which has two neutrons and a proton. No ^{4}H was known until the discovery of ^{4}H$_\Lambda$, which consists of a proton, two neutrons, and a lambda-particle.

The flood of new observations revealed no initial ordering and no systematic grouping, which moved J. Robert Oppenheimer to refer to the "nuclear zoo" in his talk at a 1956 international conference at the University of Rochester. With better measurements of particle masses and lifetimes, a semblance of order began to appear in the classification of the particles and their decay products, and the theorists brought their ingenuity to bear. A much deeper understanding of nuclear forces emerged, recognized through the award of Nobel Prizes in 1957 to T. D. Lee of Columbia University and C. N. Yang at the Institute for Advanced Study in Princeton, and in 1969 to Murray Gell-Mann of Caltech.

This exciting phase of CR research lasted only until about 1955. As the size of particle accelerators grew, so did the energies at which controlled beams of particles could be produced, and thus cosmic rays were displaced as the source of high-speed projectiles. The first results from the new generation of machines came in 1953 from the 3-GeV protons produced in the Cosmotron at Brookhaven National Laboratory on Long Island. This was followed by the 5.6-GeV protons from the Bevatron at the University of California's Radiation Laboratory in Berkeley, where beams of kaons were produced in 1955 and antiprotons were discovered in 1956.

Louis Leprince-Ringuet of the Ecole Polytechnique vividly described the coming ascendancy of accelerators at the 1953 International CR Conference in Bagneres de Bigorre (p. 290):

> . . . surgit une marée, une inondation, un déluge qui
> s'amplifie progressivement et qui nous oblige à monter de
> plus en plus haut. C'est évidemment une position qui n'est
> pas très confortable, mais n'est-ce pas là une situation ex-
> trêmement vivante et d'un merveilleux intérêt?★

Today, more than thirty years later, the inundation has long prevailed, and the field of elementary particle physics has been

★. . . an irresistible tide is rising, a flood, a deluge which is steadily growing, and we are forced to climb higher and higher. Clearly this is not a comfortable position, but is it not a situation that is extremely lively and wonderfully interesting?

transformed far beyond the imagination of any of the early workers. Not only have the accelerators increased in size—particles now travel in a vacuum around circles many kilometers in diameter—but so have the particle detectors. Rochester and Butler used a cloud chamber measuring 30 cm across; the UA-I experiment at the CERN accelerator in 1981 was several stories tall and weighed more than 200 tons. When the results of accelerator studies are published, it is not unusual to find most of the first page taken up by names of the 100-odd participating scientists and their affiliated laboratories. States and even countries vie for the selection of the next large machine, for the expected prestige and also for the economic benefits. It is a far cry from sending up a few pounds of emulsion with a balloon.

Is there any role remaining for cosmic rays in the study of high-energy collisions? Yes—perhaps even two. First, CR energies still far exceed anything remotely possible in the laboratory. The only way to examine collisions at energies above about 10^{18} eV is still by way of cosmic rays and their extensive air showers—which are very remote, extremely infrequent, and totally unpredictable in their arrival. There is also the possibility of an accidental discovery made in the course of otherwise routine experiments. A team of Bolivian and Japanese scientists studying emulsions exposed on Mount Chacaltaya have found events in which hundreds of parallel tracks appear to start from collisions outside the emulsion containers. The origin of these "Centauro" events is not yet known; they could come from the decay of a new type of particle or from a rare type of collision process.

The study of cosmic rays has made major contributions to elementary particle physics, essentially establishing it as a field of research. It is an important legacy that exemplifies the role of accident in scientific discovery and the ingenuity of scientists in improving their laboratory techniques.

Chapter 11

Footprints and Souvenirs

*L*ike much of science, the pursuit of cosmic rays is remote from everyday experience. Yet cosmic rays do create effects that touch our lives. Most of these effects are generated by protons and alpha particles, the most abundant of the primary solar and galactic rays. They penetrate both the atmosphere and solid materials, where their collisions generate new isotopes—some stable, others radioactive. Sensitive laboratory techniques have enabled research-ers to detect minute quantities of these isotopes and to use these CR products in two major applications: as clocks and as tracers. The present-day concentration of an isotope in a material can be used to determine the age of that material. The best-known example is dating by radiocarbon or carbon 14, which is produced by cosmic rays in the Earth's atmosphere and enters every living animal and plant. Measurement of ^{14}C in organic remains—cloth, skeletons, wood—has revolutionized archaeological dating. Other CR radioisotopes help to measure the ages of meteorites and the turn-over rate of lunar soil by meteorite bombardment.

Cosmic rays penetrate more than the detectors on satellites and balloons. CR can and do affect computer memories. At sea level we are continually irradiated by secondary CR particles and receive an inescapable radiation dose. Such effects tend to grow in strength as altitude increases; for instance, astronauts or passengers on high-altitude planes (such as the Concorde) might receive severe radiation doses during large solar flares. In this chapter I will discuss these effects and how some of them have been put to use.

Questions of chronology have fascinated civilized society for thousands of years. How old is the Earth, and how was it formed? How and where were humans created, or when did they evolve into recognizable human form? For a long time myths and legends provided the only answers, but the acceptance of scientific methods and the discovery of radioactivity changed this. The age of the Earth is now securely based on measurements of lead isotopes and of radioactive uranium and thorium, whose relative proportions change measurably over billions of years. History and archaeology focus on the much shorter time span within which societies developed—and left numerous relics. The dating of buildings, skeletons, and artifacts was first established from a variety of clues including historical records, chronicles of kings and dynasties, legends, and the strata of occupation at a site. When Willard Libby at the University of Chicago invented the radiocarbon dating technique in the 1940s, he initiated a revolution in historical studies.

Carbon is critical for life, for living organisms consist of a bewildering variety of carbon-based molecules. Long chains of carbon atoms, with attached loops and branches of carbon and other atoms, constitute DNA and the complex molecules needed for life and replication. The exchange of carbon dioxide between plants and animals, atmosphere and oceans is a necessary part of the life cycle. Carbon is synthesized in stars; the carbon we find on Earth consists almost entirely of a mixture of two isotopes: 98.9 percent is ^{12}C, six protons and six neutrons in each nucleus, and the other 1.1 percent is ^{13}C, which carries an additional neutron in each nucleus. Carbon isotopes with half-lives of minutes or shorter have been created in laboratories but hold no geological or historical interest. Carbon 14, which is generated by cosmic rays in the Earth's atmosphere, is quite different.

As primary CR penetrate the atmophere, they encounter air of increasing density, and by 50,000 feet in altitude most will have undergone nuclear collisions in which neutrons are produced. The neutrons quickly lose enough energy to be "thermalized," and their speeds are reduced to the point where they can be captured by nitrogen nuclei to produce ^{14}C.

The newly created ^{14}C is radioactive, with a half-life of 5,730 years. It decays back to nitrogen after emitting an electron. The radioactivity does not affect the chemical ability of the carbon 14 to combine with other atoms to form molecules. The radiocarbon combines with oxygen and produces radioactive carbon dioxide, $^{14}CO_2$, which then spreads through the biosphere. Many of these molecules attach to minute dust particles and follow air currents, becoming incorporated in rain. Some carbon dioxide is absorbed by plants that are eaten by humans and other animals; some goes into rivers and then into oceans. What makes all this possible is the long half-life of carbon 14, far longer than the typical seven years needed to spread from cosmic rays in the stratosphere to the atmosphere, weather systems, plants and animals, and the oceans.

Carbon 14 has been produced long enough to result in a steady inventory on Earth, with production just balanced by decay. There are roughly 54 tons of radiocarbon on Earth, with an average concentration of one carbon-14 atom for every trillion regular carbon atoms. This is the concentration existing in all living things. When a plant or an animal dies, its intake of carbon dioxide with its carbon 14 ceases. Thereafter, the $^{14}C/^{12+13}C$ ratio changes slowly as carbon 14 decays. Determining this ratio enables us to estimate the age of the wood, skeleton, paper, or cloth.

Measuring such low concentrations of carbon 14 is not a simple process. In the first method, devised by Libby, electrons from the carbon decay in a small sample are counted. Detecting the electrons requires a combination of particle counters to distinguish them from the larger numbers of particles coming from local radioactivity and sea-level cosmic rays. Because the objects to be dated are usually unique and testing destroys any sample taken, there is understandably great pressure to keep the sample sizes small; one-tenth of a gram (about 1/300 ounce) is usually the largest available. This effectively sets a limit to the age that can be determined in this way, for in very old materials most of the radiocarbon will have decayed, and the counting rate from a tiny sample will be lower than background levels.

The radioactive signal of the equilibrium concentration of radiocarbon shows up in the emission of 13.5 electrons per minute for each gram of carbon in a fresh sample. In wood that is 2,000 years old, for example, about 79 percent of the radiocarbon will have survived and the decay rate will be about 10 counts per minute. By 4,000 years, the corresponding rate will be nearly 8 counts per minute.

A far more sensitive technique for carbon-14 detection was developed in the 1970s; it identifies atoms by their masses rather than by their radioactivity. A minute sample is vaporized and the fully ionized atoms sent through a cyclotron or other type of accelerator. The high-speed nuclei emerge to travel through well-calibrated magnetic fields in which the curvature of their paths depends on their speed, mass, and charge. Counters, similar to the ones used in CR work, measure their energies and speeds, and the arrival of each nucleus is individually noted. Only a few milligrams of sample are needed, and the carbon 14 can be clearly separated from the more abundant carbon 12 and 13.

Converting measured carbon-14 concentrations to ages is not a straightforward process. "Dates" derived from radiocarbon measurements must be corrected for variations in the CR's production rate. There will also be variations in the diffusion rate of the radiocarbon. Still, with the most sensitive techniques, ages of up to 70,000 years can now be determined under favorable circumstances.

From measurements of the radiocarbon in the annual rings of bristlecone pines and sequoia trees, radiocarbon calibration has been constructed and the basic technique validated. Growth patterns exhibit a long-term drift between carbon and "true" dates, with additional short-term variations. Wood samples whose rings are no more than 3,000 years old tend to have ages overestimated by the radiocarbon technique, while the effect is reversed for older samples. The size of this effect is generally in the 10 to 15 percent range; thus wood whose rings are 7,000 years old will be seen, by radiocarbon, to be about 600 years younger. One cause of this difference is the slow drift in the strength of the Earth's magnetic field, which controls the entry of low-energy CR that produce the radiocarbon. Another factor is the variation of CR intensity during the 11-year solar cycle. Even with the calibration from tree rings, caution is still needed in interpreting the measurements. For example, apparent ages of 2,000 years have been found for some living organisms whose carbon dioxide had come from limestone on the floor of a freshwater lake, which, in turn, was deficient in radiocarbon.

Serious problems affecting radiocarbon dating have been created by humans. The burning of large quantities of fossil fuels, begun in the nineteenth century, has injected enough low-^{14}C carbon into the atmosphere to decrease the radiocarbon ratio by about 3 percent. Even worse, before the atmospheric testing of nuclear weapons ceased in the early 1960s, so much radiocarbon had been produced and dispersed that the atmospheric inventory was doubled. The overall result of these alterations and various other uncertainties is that future calibrations cannot be based on today's wood, and dates within the past 200 years are hard to establish with reliability. But the tree-ring calibration does still show that, to within better than 50 percent, CR intensity has been fairly constant over the past 8,000 years, and in general the radiocarbon method is still of great use.

The tracking of radiocarbon is effectively confined to the Earth because of the atmosphere and the life cycle. Many other radioisotopes are produced by both galactic and solar CR, on the Earth and on the moon, in sufficient quantities to be of interest. The longest-lived of these are listed in Table 11.1. Beryllium 10 (^{10}Be) is particularly interesting because of its use as a chronometer for CR travel in the galaxy. Most ^{10}Be in the solar system is thought to have been produced by cosmic rays, some in space and rather less by collisions in our atmosphere. Its half-life is long enough for it to be transported from the atmosphere to the surface, where it is found in sea water, polar ice, igneous rocks, ocean sediments, and nodules on the ocean floor. Another isotope produced by cosmic rays, ^{26}Al, with a half-life close to three-quarters of a million years, has chemical behavior similar to that of beryllium. The Be/Al ratio

Table 11.1 *Long-lived isotopes produced by cosmic rays*

Isotope	Half-life (years)
^{10}Be	1.6×10^6
^{26}Al	7.2×10^5
^{36}Cl	3.0×10^5
^{40}K	1.3×10^9
^{53}Mn	3.7×10^6
^{59}Ni	5,272
^{129}I	1.6×10^7

is useful as an index for determining how fast atmospherically produced isotopes are carried through geological processes. In addition, their measured abundances show that the cosmic rays that generated them must have had roughly the same intensity for the past 5 to 10 million years.

On the Earth, wind, rain, and vegetation combine to eliminate or mask the evidence of old meteorite impacts, but on the moon those surface features last indefinitely—or until the arrival of another meteorite. The samples retrieved by the Apollo astronauts during the 1969–1972 missions established an absolute chronology for some lunar surface features. Before that time only a relative chronology could be constructed; for example, in a place where debris from one crater lay across another feature (perhaps another crater), we knew the sequence of formation but not how long ago the events had occurred. Now, by using radiodating methods on lunar samples, we know that the moon's age is about 4.6 billion years and that most of the major cratering ceased by 3 billion years ago. Since then there have been occasional violent impacts with the arrival of large bodies and a steady drizzle of much smaller objects, ranging in size from small rocks through micrometeorites of millimeters or smaller down to individual nuclei from the solar wind and from solar and galactic CR.

Small meteorites can make microscopic pits in lunar rocks; larger ones carve small craters in the lunar surface and throw out debris. Because of the low surface gravity on the moon, this debris will travel farther than it would on Earth. The lunar surface is thus continually being churned up, with fresh debris overlying that from earlier impacts. The net result is a "gardening," with deeper layers turning over less rapidly because they require the infrequent arrival of larger meteorites.

With no shielding atmosphere to impede them, primary CR penetrate the lunar surface and produce radioisotopes, as they do on Earth. CR penetration naturally decreases with depth, so material closer to the surface receives a larger dose of cosmic rays than does the deeper material. To this must be added the gardening effects, in which fresh (less irradiated) material is brought to the top by cratering, where it is exposed to a larger radiation dose than it otherwise would have received.

In addition to producing radioisotopes, cosmic rays are responsible for causing radiation damage on a different scale, within certain types of crystals. Some energy lost by cosmic rays is stored within the crystals and can be retained for a long time. Subsequent heating of the crystals releases the stored energy, which appears as light. This *thermoluminescence* (TL) reveals the radiation dose that the crystal received and thus the number of CR particles that went through the crystal. TL monitors are widely used as industrial dosimeters, routinely worn by people who work near X-rays and radioactive materials.

During some Apollo missions, astronauts hammered hollow tubes more than 2 meters into the lunar surface to collect core samples that preserved the relative positions of the different layers. From the TL in these cores, the CR exposure at different depths and thus the gardening history could be calculated. In this way, it was found that the top half-millimeter of the lunar surface is turned over 100 times in a million years. At a depth of 1 cm, 50 pecent of the material is untouched in that period of time; it takes about 10 million years to turn over the top centimeter. The turnover time scale lengthens to a billion years between 10 cm and 100 cm.

Measurement of TL, even in small crystals, demonstrates a bulk effect in that the energy is accumulated throughout each crystal from the passage of many cosmic rays over many years. The effect of an individual CR particle is not measured. Tracks of individual cosmic rays have been seen, however, through the process of chemically etching suitable materials. Figure 11.1 shows such tracks, which were revealed by etching in sodium hydroxide. During the Apollo missions, some detectors were deployed on the moon's surface and promptly brought back. One of the more striking demonstrations of CR tracks was seen in a filter from the camera on the Surveyor 3 lunar lander. This unmanned vehicle had landed on the Moon in 1967, and the filter was retrieved by the Apollo 12 mission over two years later.

When the solar system formed by the contraction of a great cloud of gas and dust, most of the condensed mass ended up as the sun, with less than 1 percent as much in the planets and their satellites. It was thought that most of the original cloud was somehow expelled from the solar system. A tiny fraction of the original mass remains in the form of comets, meteorites, and dust. For many years, comets could be studied only by remote observation of their often spectacular but generally unpredictable appearances. A first step

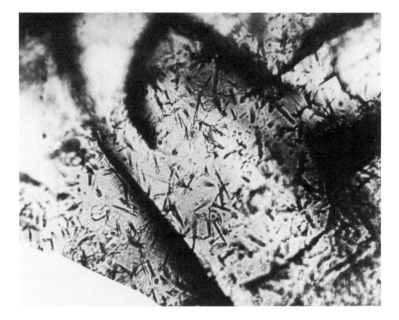

Figure 11.1
Tracks of cosmic rays, revealed after etching a crystal from a lunar rock. Each of these tracks is less than 1/100 mm in length. (Photograph courtesy of Robert M. Walker, Washington University.)

toward laboratory-based experiments was taken when the 1985 and 1986 space probes intercepted the periodic comets Halley and Giacobini-Zinner, whose returns can be accurately forecast. Plans are being made to intercept future comets, collect samples, and return them to Earth.

The study of interplanetary dust started with Earth-based observations of the *gegenschein* and the *zodiacal light,* sunlight reflected from dusty regions in the solar system. From spectroscopic analysis we know that submillimeter dust grains are responsible for these phenomena and that the solar system contains a great many other dust grains, for example from comet tails. Dust particles have been collected by high-flying airplanes and identified, by their composition, as certainly not being terrestrial in origin. These interplanetary dust particles (IDPs), typically a few hundredths of a millimeter in diameter and weighing around one-billionth of a gram, have been examined with a battery of laboratory techniques at the very limits of sensitivity. Beyond showing that the composition does not match with terrestrial material, the measurements have revealed a record of CR irradiation (Figure 11.2).

In many of the particles, CR tracks have been observed in the electron microscope after chemical etching. Track densities as high as 10^{11} tracks per square centimeter have been seen, for example, in a grain of the mineral olivine ($MgFeSiO_4$). With present-day CR intensities, it would take only about 10,000 years to accumulate this number of tracks. Observation of these tracks also tells us that

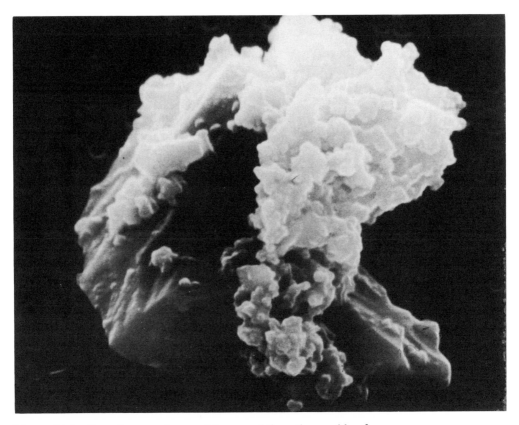

Figure 11.2 *Interplanetary dust particle, around three-thousandths of a centimeter in diameter. Measurement of the isotopic composition of hydrogen in these particles has demonstrated their extraterrestrial nature and also provides information on their bombardment by CR during their interplanetary travels. Tracks have been found in some of these particles. (Photograph courtesy of Robert M. Walker, Washington University.)*

during interplanetary travel and even atmospheric entry, the grains could not have been heated above roughly 500 °C, for at higher temperatures the tracks would be obliterated.

Another indicator of the extraterrestrial origin of the grains is the detection of relatively large quantities of those isotopes of the rare gases helium, neon, argon, krypton, and xenon that can be produced only by CR interactions. Many techniques being applied to the IDPs were developed over the years as the status of more easily obtained meteorites shifted from museum curiosities to prized scientific specimens.

Ages for many meteorites have been determined by applying radiometric techniques using isotopes of lead and uranium and an isotope of rubidium that has a half-life of 49 billion years. This method was previously used for dating terrestrial and lunar samples. "Age," in this context, means the time since the meteorite solidified enough to retain the gaseous radioactive daughter products that would otherwise leak away. In this way, ages up to 4.6 billion years have been deduced.

Meteorite "CR ages" have also been found from analyzing CR tracks in mineral grains, assuming that the tracks accumulated throughout the time when the meteorite surface was exposed to cosmic rays at the present-day intensity. CR exposure ages of 5 to 50 million years have been found for some stony meteorites, and rather greater ages (200 to 1,000 million years) for some iron meteorites. These ages correlate with the mean lifetimes for these types of meteorites in orbits that cross those of the Earth and Mars; the CR ages then suggest that these objects might have originated in the breakup of a group of larger bodies millions of years ago. The picture is one of steady meteorite generation: fresh surfaces are formed and then accumulate CR tracks over times quite short relative to the age of the solar system.

Other isotopes, such as ^{39}Ar and ^{36}Cl, are produced by cosmic rays inside the meteorite after it has landed on the Earth's surface. Terrestrial "ages" (since impact) from 270 years to 1.5 million years have been measured by using these isotopes.

Roentgen's dramatic picture of the bones in his hand immediately caught the attention of the public and the medical profession. Since then, for nearly a century, X-rays have been used for diagnostic purposes and have become an indispensable tool of modern medicine. At one time X-ray machines were widely used in shoe stores, and children enjoyed seeing the skeletal images of their wiggling toes as shoes were being fitted. X-rays have also been used therapeutically, especially in the treatment of some types of cancer. If these were the only ways in which the word *radiation* turned up in everyday use, we might well have few worries, but Hiroshima and Nagasaki changed that. To what extent do cosmic rays produce unwanted and adverse effects?

Many physical processes used to detect cosmic rays also occur when CR and other high-speed particles travel through living tissue. Some of the particles' kinetic energy is transferred to the tissue, where electrons can be ejected from atoms. In this way chemical bonds can be broken and the disrupted molecules can serve as the focus for cancer to develop. Radiation is used to kill cancer cells, but healthy cells can also be killed.

Different systems of units describe radiation doses. In the older system, the *rem* was defined to correspond to the absorption of around 100 ergs/g of energy in tissue. (The qualification "around" refers to the differing biological effectiveness of different forms of radiation such as protons, neutrons, X-rays, and electrons, which are included in a more specific definition.) By international agreement, a newer unit, the *sievert* (sv) has been adopted, with 1 sv equal to 100 rem. Because so much of the popular literature still uses the older units, I will do so as well.

The average radiation dose that every person receives each year is about one-tenth of a rem, or 100 millirem (100 mr), derived from three roughly equal sources. The contribution of the cosmic rays varies somewhat with geographical latitude and longitude; this radiation dose is distributed evenly throughout the body. A second contribution, delivered internally and concentrated in some organs, comes from the radioactive potassium 40 that is part of our natural chemical makeup. The third source is the average daily medical dose (X-rays and nuclear medicine) received by the population, at least in developed countries. Therapeutic medical doses will generally be much larger, but those are not received by the general population and are applied only when there is a clear medical reason and when benefits outweigh risks.

Inhalation of radon, a gaseous daughter product from radium, may expose many people to an additional 100 mr, mainly in the lungs, but this dose varies widely across the country, depending on factors such as local geology, type of building material, and degree of ventilation in housing. The potential hazard posed by excessive radon inhalation is widely recognized, and commercial kits can be bought for easy monitoring.

A lethal dose of radiation, such as was received in 1945 by people in Hiroshima and Nagasaki, exceeds 400 rem, but the CR dose rate for an individual, even accumulated over a lifetime, is far lower. On the other hand, the total CR dose received annually by the entire U.S. population is 240 million $\times$ 30 mr = 7.2 million rem. The incidence of cancer is in the range of one case for every 10,000 rem; so perhaps 720 cases of cancer are unavoidably induced each year in the United States by cosmic rays. This is far lower than the total annual cancer rate and thus should not be a cause for great concern—nor is there anything we can do about it.

Potentially more serious is the dose that could be received by airplane passengers and crew and astronauts who are traveling at high enough altitudes to be exposed to large numbers of solar CR during major flares. The giant flare of February 1956 would have produced a peak dose rate of 100 mr per hour in people flying at 35,000 feet, far above the usual commercial airplane altitudes of those days but quite normal today. During a single transatlantic flight, a passenger could thus receive a dose at least equal to the normal yearly accumulation. Crew members in a supersonic plane, traveling at around 60,000 feet for 500 hours each year, will accumulate a dose of 500 mr from galactic CR. This amount is well above the normal population dose and is equal to the recommended annual maximum for industrial radiation workers. To avoid large doses during a solar flare, an airplane would have to reduce altitude rapidly and gain protection from the greater thickness of overlying atmosphere. This defensive maneuver is not available to astronauts, who could encounter additional doses in passing through the Van Allen radiation belts. In the center of the belts, a dose rate of 20 rem per hour could be received. This can be avoided by keeping spacecraft orbits below 400 km or above 6,000 km.

The unseen hazard presented to astronauts by cosmic rays has been vividly illustrated in two ways. During the Earth-moon stages of their journeys, several Apollo astronauts reported seeing bright flashes of light, even when their eyes were closed. The flashes were described as "pinpoints that disintegrated rapidly" and "frequently streaks, with occasional double flashes." The nature and frequency of these flashes are consistent with their production by cosmic rays. Some of these CR (probably heavily charged particles) produced Čerenkov radiation in the eyeball, while others might have hit the retina directly.

There is no way that the astronauts' eyes can be examined for residual evidence of CR traversal, but support for this explanation of the flashes comes from particle tracks in the plastic helmets worn by the astronauts. As shown in Figure 11.3, etching the plastic has revealed tracks from heavy CR, identified by their track dimensions.

If cosmic rays can cause light flashes in astronauts' eyes, they can also effect changes in the brain, but we have no evidence yet for harmful effects from single particles. Computer memories might be similarly affected, and the chance for this increases during long-term satellite flights. Some types of computer problems are termed *soft fails,* in which a single digit changes without command, and testing reveals no way of duplicating the event. In contrast, *hard fails* are the result of faulty but replaceable components. The passage of a CR particle through a computer chip can trigger soft fails, and the question is whether their frequency poses a hazard. The conclusion drawn from several studies is that only heavily ionizing particles generally represent a potential threat, but the problem will become more serious as more compact computer devices come into use. It is possible, however, to incorporate error-correcting codes to guard against random errors so that computer instructions are not incorrectly interpreted.

There is a recurrent theme in scientific discovery, well described by the aphorism "one person's noise is another's signal." What is an intrusive effect in some experiments may turn out to be of great

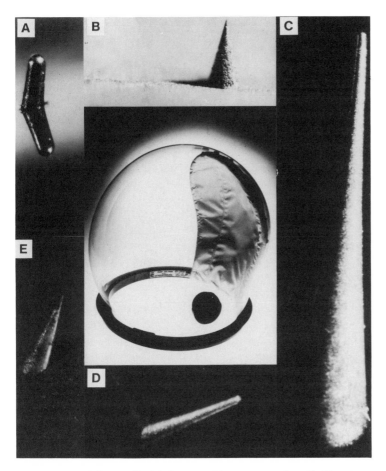

Figure 11.3 *The Apollo 8 helmet used by astronaut Lovell. Cosmic ray tracks in this and other plastic helmets were revealed by chemical etching, and replicas are shown in the surrounding photographs. Tracks A, D, and E were found in the helmets of Apollo 12 astronauts Conrad and Gordon; tracks B and C were in a control helmet exposed to primary CR. Track C was 0.7 mm in length, and the others were all in the range 0.35–0.60 mm. (Photograph courtesy of R. L. Fleischer et al.,* Science *170, 11 December 1970, 1189–1191, © 1970 by A.A.A.S.)*

importance in its own right. The discovery of cosmic rays was accidental, through an attempt to understand an effect (electric charge leakage) that had emerged as a nuisance during experiments with radioactivity. Establishing the extraterrestrial nature of the new radiation and deciding among competing interpretations of the data took close to twenty-five years, largely because of the unrecognized role of the Earth's magnetic field and the complexity of the secondary effects in the atmosphere. Indeed, those effects were more than just complex; they resulted from the interplay of processes not even remotely imagined at the time.

The further accidental discovery of the positron and mesotron opened up the field of elementary particle physics. For twenty years, until the construction of the large accelerators in the 1950s, new elementary particles could only be found through cosmic rays—always, of course, by accident. As complementary information came flooding in from optical and radio astronomy, the 1948 discovery of the heavy primary particles broadened the astrophysical base that remains the central theme of cosmic ray studies. With the advent of X-ray and gamma-ray astronomy, which required very high altitude observations, we entered the era of high-energy astrophysics, the current and more comprehensive designation. Along the way progress was made through the symbiotic development of cosmic ray physics, new types of particle and radiation detectors, the theory of particles and their interactions, and the launching of the space age.

The scope of cosmic ray physics has continued to widen. A survey of the annual number of scientific publications on cosmic rays shows the following figures: in 1924 there were 9 papers published on cosmic rays; in 1933 there were 184; in 1950 there were close to 400. In 1960 and 1970 the numbers increased to approximately 500 in each of those years, and in 1986 there were about 750 papers on cosmic rays. We do not expect the increase to continue at this pace, however. In spite of its impressive record, the future of CR research is very cloudy. Yet there are still important measurements waiting to be made. For example, more data are needed so that the fine details of the charge spectrum can be examined and the abundances of individual isotopes can be determined. The measurement of the energy spectrum, especially at the highest energies and of the high-Z nuclei, must continue so that researchers can construct more refined models for the origin and propagation of cosmic rays. Further searches are needed to pin down the abundances of the antinuclei or to set upper limits. Many observations require space-borne hardware, but no major CR satellite will fly soon. On the ground, research on extensive air showers will continue, and there will probably be a modest amount of ballooning, for its own interest and also for testing prototype experiments that can later go up on satellites.

The history of science has shown how research fields flourish and then dwindle as new fields emerge. It may be that the contributions of cosmic ray research have for the present been exhausted. Those of us who have had the good fortune to participate in this research have had so much enjoyment that we should not begrudge others the excitement of seeing their fields prosper and move ahead. It is this sense of novelty and continuity, consolidation and discovery that makes scientific research such a wonderful enterprise.

The most fitting commentary on cosmic rays was made by J. J. Thomson and G. P. Thomson in 1928, long before the full range of this exciting subject was known. In *Conduction of Electricity through Gases,* they wrote (p. 12): "It would be one of the romances of science if these obscure and prosaic minute leakages of electricity from well-insulated bodies should be the means by which the most fundamental problems in the evolution of the cosmos had to be investigated."

Works Cited

Baltrusaitis, R. M., et al. 1985. *Nuclear Instruments and Methods,* vol. A240.

Drake, S., trans. 1957. *Discoveries and Opinions of Galileo.* Garden City, N.Y.: Doubleday and Company.

Leprince-Ringuet, L. 1953. "International Congress on the Cosmic Radiation." Mimeo, University of Toulouse.

Millikan, R. A. 1926. *Physical Review* 27:360.

Störmer, C. 1955. *The Polar Aurora.* Oxford: Clarendon Press.

Taylor, A. M. 1967. *Imagination and the Growth of Science.* New York: Schocken Books.

Thomson, J. J., and G. P. Thomson. 1928. *Conduction of Electricity through Gases,* 3rd ed. Cambridge: Cambridge University Press.

Thrower, N. J., ed. 1981. *The Three Voyages of Edmond Halley in the Paramore.* London: The Hakluyt Society.

Young, C. A. 1898. *The Sun.* New York: D. Appleton and Company.

Bibliographical Note

Most of the literature on cosmic rays is in research journals or monographs and is accordingly technical. There have been several more popular treatments, but these are now rather dated, either through having been bypassed by the flood of information in recent years or by the shifting balance of interests within cosmic ray physics. Among the older popularizations, two are definitely worthy of mention, although both are out of print. The first is *Cosmic Rays* by Bruno Rossi (New York: McGraw-Hill, 1964). This book is a classic by one of the cosmic ray pioneers. Rossi has made major contributions to the field since 1924, and the High Energy Astrophysics Division of the American Astronomical Society now awards an annual Rossi Prize. The second book is Martin A. Pomerantz's *Cosmic Rays* (New York: Van Nostrand Reinhold, 1971), an excellent survey. Pomerantz's research has been mostly concerned with the solar-terrestrial aspects of cosmic rays, and his book is particularly strong in its treatment of these topics.

The discovery and early studies of cosmic rays are included in a popular history of physics by Emelio Segre, *From Atoms to Quarks* (San Francisco: W. H. Freeman, 1980). Segre won a Nobel Prize for the first observations (in 1956) of antiprotons, produced at the University of California's Bevatron. A similar treatment, but with many more illustrations, is *The Particle Explosion* by Frank Close, Michael Marten, and Christine Sutton (New York: Oxford University Press, 1987). This book is one of the best popular surveys of the history of particle physics.

A few other books deserve mention. *The Early History of Cosmic Rays,* edited by Y. Sekido and H. Elliot (Dordrecht: Reidel, 1985) contains personal recollections by many of the early researchers. *The Birth of Particle Physics,* edited by Laurie M. Brown and Lillian Hoddeson (Cambridge: Cambridge University Press, 1983) contains papers presented at a conference and includes recollections from several of the pioneers of CR particle physics, including Carl Anderson, Bruno Rossi, and Marcello Conversi. Another set of recollections appears in the proceedings of a conference held in 1987 to celebrate the fortieth anniversary of the discoveries of pions and V-particles. *40 Years of Particle Physics* is edited by B. Foster and P. H. Fowler (Bristol and Philadelphia: Adam Hilger, 1988). An extended treatment of particle history, with considerable theoretical detail, is given by Abraham Pais in *Inward Bound* (Oxford and New York: Oxford University Press, 1986). In *Origins of Magnetospheric Physics* (Washington, D.C.: Smithsonian Institution Press, 1983) James A. van Allen provides an account of his discoveries made with the use of rockets and satellites.

Articles on cosmic rays appear in *Scientific American* from time to time and can be located by consulting the cumulative index. Useful compilations also appear in *Mercury,* a popular magazine published by the Astronomical Society of the Pacific.

Index